Naturalists' Handbooks 35

Rock pools

PETER J. HAYWARD

Pelagic Publishing

First published in 2022 by
Pelagic Publishing
20–22 Wenlock Road
London N1 7GU, UK

www.pelagicpublishing.com

British Library Cataloguing in Publication Data
A catalogue record for this book is available from the British Library

ISBN 978-1-78427-359-0 Pbk
ISBN 978-1-78427-360-6 ePub
ISBN 978-1-78427-361-3 PDF

https://doi.org/10.53061/GXHH2122

Cover images: Tompot Blenny *Parablennius gattorugine* © Alex Mustard/naturepl.com; Snakelocks Anemones *Anemonia viridis*; nudibranch *Doto fragilis* © J.S. Ryland.

Contents

Editor's preface v

Acknowledgements vi

About the author vi

About Naturalists' Handbooks vii

1. **Introduction** 1
2. **The pool environment** 5
3. **Rock-pool seaweeds** 15
4. **Rock-pool animals** 35
5. **Identification** 54
   - *Key A Guide to major invertebrate animal groups* 55
   - *Key B Sessile, modular animals* 59
   - *Key C Sea spiders (Pycnogonida)* 67
   - *Key D Isopods* 70
   - *Key E Amphipods* 75
   - *Key F Decapods* 93
   - *Key G Shelled gastropods* 99
   - *Key H Chitons* 108
   - *Key I Bivalves* 110
   - *Key J Polychaetes* 114
   - *Key K Heterobranchia* 121
   - *Key L Sea anemones* 128
   - *Key M Echinoderms* 136
   - *Key N Rock-pool fishes* 141
6. **Investigating rock pools** 151
7. **References and further reading** 160

Index 168

Editor's preface

Many of us will have first encountered rock pools as children on seaside holidays or as students on marine biology field courses. They offer easy access to a different world, full of unfamiliar and fascinating creatures, such as hermit crabs, skeleton shrimps, starfish, sea anemones and sea squirts. Rock pools are remarkable for the rich diversity of invertebrate groups that are present, most of which cannot be seen inland. Although rock pools may appear at first glance to be simple habitats, they vary enormously in factors such as temperature, salinity and oxygen content. They can be extreme environments and their inhabitants have some remarkable adaptations in order to survive there. There are also many curious and little-understood associations between species. There is much waiting to be found out about rock pools and the species that live in them.

This *Naturalists' Handbook* provides an introduction to the natural history, biology and ecology of the animals and plants in rock pools, and advice on how to study them, together with identification keys to the major animal groups.

I hope that this book will encourage people to study rock pools. They offer convenient microcosms in which to investigate the biology and ecology of many fascinating species.

This *Naturalists' Handbook* on rock pools complements other titles in the series about the seashore: *Animals on seaweed* (No. 9), *Animals of sandy shores* (No. 21) and *Snails on rocky sea shores* (No. 30).

William D.J. Kirk
January 2022

Acknowledgements

Sources of the data plotted in Figures 2.4–2.7, 3.10, 3.11, 3.24, 4.5, 4.6 and 4.18–4.20 are noted in the figure legends and cited in the list of references and further reading (Section 7). A small number of diagrams and drawings reproduced or redrawn from published sources are similarly acknowledged. The majority of the drawings and diagrams accompanying the identification keys were originally prepared for Naturalists' Handbook 9 (*Animals on seaweed*: Hayward 1988), while others have been reproduced from the *Handbook of the Marine Fauna of North-West Europe* (Hayward & Ryland 1995, 2017). Of the latter, a number were originally the work of the illustrators P.J. Llewelyn (Figures D.4, D.9, F.2–F.12, F.14–F.16, L.6, L.9, L.12, M.2–M.4, N.2, N.3, N.4b, N.5b, N.7–N.10, N.11a and N.13–N.21), Nigel Gerke (Figures G.10, G.16–G.19 and G.29) and Toni Hargreaves (Figure M.7). I wish to thank those friends and colleagues whose published researches I have consulted, and cited, in the preparation of this work, while accepting responsibility for any errors of fact or interpretation. I am grateful to Prof. John S. Ryland, Dr Joanne S. Porter, Dr Jim R. Ellis and Dr Peter E.J. Dyrynda, who kindly provided a number of colour images, as acknowledged in the figure captions. Sam and Ben Fenwick provided valuable assistance in the field.

About the author

Peter J. Hayward DSc. began his career as Scientific Assistant at the Natural History Museum, London, and retired as Senior Lecturer in Marine Biology at Swansea University. Marine invertebrates have been his abiding interest, through many spring and autumn field courses, and in practical laboratory exercises. He has published several books on marine biological topics, along with numerous scientific papers, especially on the marine Bryozoa. He has served as co-editor of the *Journal of Natural History* and editor of the *Zoological Journal of the Linnean Society*, and has been a frequent contributor to *BBC Wildlife*.

About Naturalists' Handbooks

Naturalists' Handbooks encourage and enable those interested in natural history to undertake field study, make accurate identifications and offer original contributions to research. A typical reader may be studying natural history at sixth-form or undergraduate level, carrying out species/ habitat surveys as an ecological consultant, undertaking academic research or simply developing a deeper understanding of natural history.

History of the Naturalists' Handbooks series

The *Naturalists' Handbooks* series was first published by Cambridge University Press, then Richmond Publishing and then the Company of Biologists. In 2010 Pelagic Publishing began to publish new books in the series, along with updated editions of popular titles such as *Bumblebees* and *Ladybirds*. If you are interested in writing a book in this series, or have a suggestion for a good title please contact the series editor.

About Pelagic Publishing

We publish scientific books to the highest editorial standards in all life science disciplines, with a particular focus on ecology, conservation and environment. Pelagic Publishing produces books that set new benchmarks, share advances in research methods and encourage and inform wildlife investigation for all. If you are interested in publishing with Pelagic please contact editor@pelagicpublishing.com with a synopsis of your book, a brief history of your previous written work and a statement describing the impact you would like your book to have on readers.

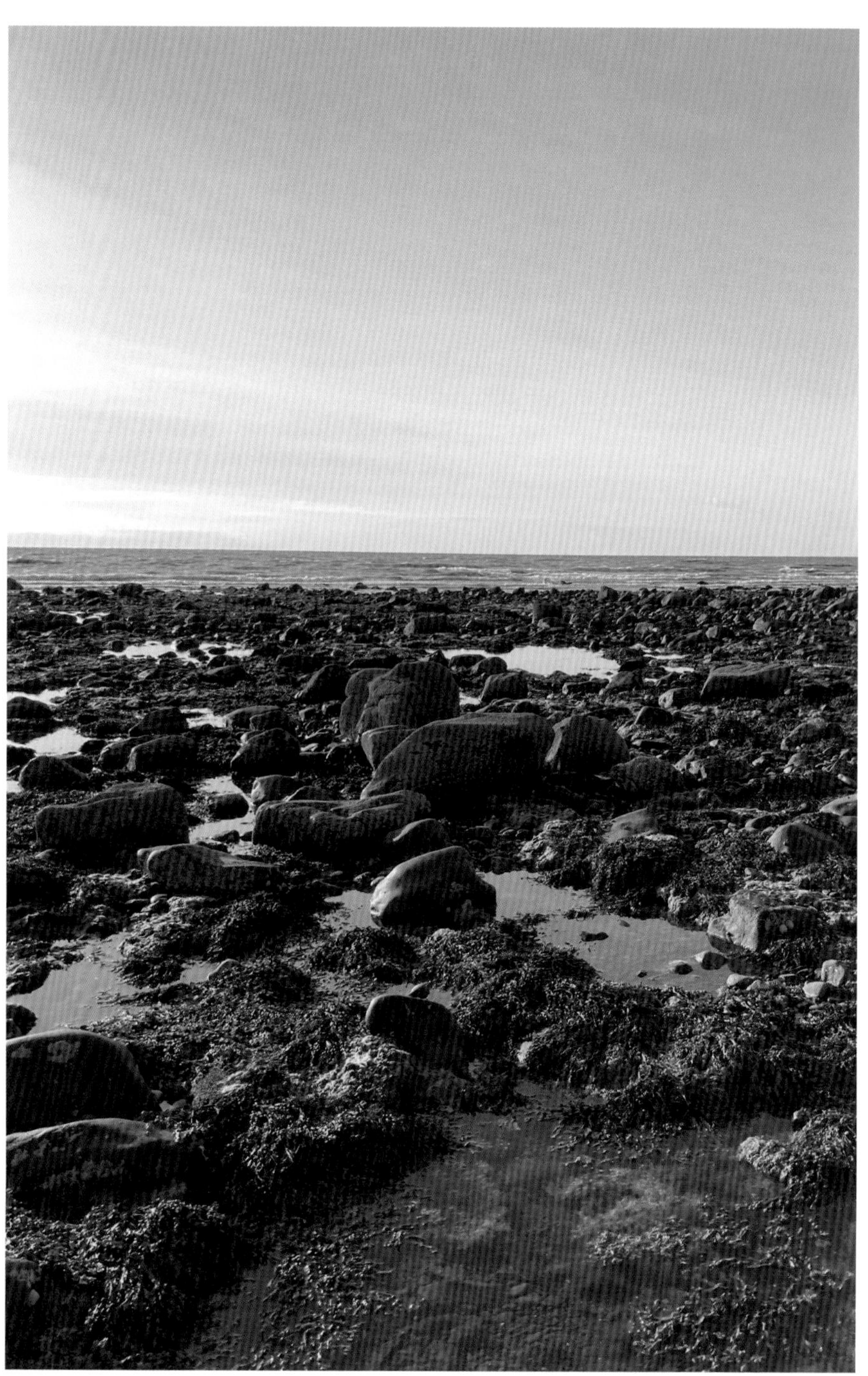

Low tide on a rocky shore: Aberarth (Dolauarth), West Wales. (Photo: David Hawkins)

1 Introduction

On rocky coastlines, receding tides may leave standing pools – rock pools or tide pools – that have long held a fascination for the amateur seashore naturalist, revealing glimpses of colourful and curious marine plants and animals. Victorian naturalists, such as P.H. Gosse and G.H. Lewes, first popularised 'rockpooling', and stimulated an interest in marine aquaria in which to observe the form and behaviour of rock-pool inhabitants for longer than a brief intertidal period. The huge marine-life centres now located in many coastal cities have mostly eclipsed small local aquaria and offer spectacular displays of marine organisms, usually from the broadest range of environments and habitats, worldwide. Moreover, rock-pool ecology remains a frequent component of biological science courses, at all educational levels, and rock pools and rock-pool studies remain the focus of an enthusiastic body of amateur naturalists.

Field courses provide essential practical experience in biological diversity and basic ecology for all biological sciences students, and the traditional marine field course has always been especially important for aspiring zoologists. Students of botany encounter the richest plant diversity in terrestrial ecosystems, but animal diversity is far greater in the sea than in terrestrial or freshwater habitats, which tend to be dominated by relatively few of the largest taxonomic groups – phyla – of macroscopic organisms. Diptera (flies) and Coleoptera (beetles) will constitute the overwhelming majority of the fauna, in terms of numbers of species, numbers of individuals, and total bulk, or biomass. A modest diversity of nematodes, annelid worms, crustaceans and molluscs completes the variety of macroscopic invertebrate phyla present in temperate freshwater and terrestrial habitats.

Coastal marine ecosystems in the temperate northeast Atlantic region do not support the huge numbers of species found in field, fen and forest – the British Isles has a recorded beetle fauna of around 4,000 species, for instance – but the number of phyla to be found, the phyletic range of coastal marine faunas, is very much greater. Representative species of 19 phyla of macroscopic invertebrate animals might be expected on a comprehensive marine field course, encompassing all seashore habitats and a variety of physical environmental regimes. Sedimentary shores – from mud, through increasingly coarse grades of sand, to gravels and cobbles – often have the lowest diversity in terms of species, although in muds and fine sands on sheltered seashores a few very abundant species may constitute a huge biomass. The species richness of some sand beaches is not always evident, most organisms living burrowed into the sediment, and sheltered fine-sand beaches may be populated by numerous species of annelid worm, small crustacean, mollusc and echinoderm (e.g. starfish), and some, such as the lugworm *Arenicola marina* and the thin-shelled bivalve *Tellina tenuis*, achieve dense populations. Rocky shores are not subject to the storm-driven seasonal disturbance that affects mobile, sedimentary shores. Rock is permanent but erodes to provide a much more heterogeneous environment than that of the sandy shore, and consequently a far greater variety of micro-habitats, and a greater species richness. Rock also provides firm attachment for seaweed populations and, thereby, a further range of microhabitats for colonisation by animals.

This *Naturalists' Handbook* is a guide to rock-pool habitats and inhabitants, and an introduction to the ecology of rock-pool environments, applicable to the rocky coasts of the British Isles and Ireland, and the Channel coasts of mainland Europe. The twice-daily period of tidal retreat – termed 'emersion' – is the most persistent stress imposed on all intertidal plant and animal communities of north-west Europe. The sessile fauna of rocky seashores – barnacles, limpets, anemones – closes up as the tide retreats, while slow-moving sedentary animals – winkles, topshells, dog whelks – withdraw into damp, shaded shelter. Most of the mobile fauna must either move down shore with the tide, or retreat beneath seaweeds or into standing pools of water. Rock pools supply refuge during low-tide periods for free-ranging decapods (e.g. crabs) and fish, and the diversity and density of the pool community may thus fluctuate between successive low tides: some species, especially hermit crabs and some littoral fish, appear to lead a peripatetic existence, moving up, down and along shores, between tides, and through tidal and seasonal cycles.

There are, however, also rock pool residents, both temporary and permanent, as well as other species that occur only sporadically, and might be regarded as what an ornithologist would term 'accidental'. Larvae of rock limpets (*Patella depressa, Patella ulyssiponensis* and *Patella vulgata*), and the epiphytic Blue-rayed Limpet *Patella pellucida* settle and metamorphose on encrusting coralline algae in lower-shore rock pools, where they remain until they have grown to a safe size, at which the risks from predation and desiccation are sufficiently reduced that the juveniles may leave the pool

epiphytic
living on or attached to a plant

nurseries and migrate to their adult habitats. Permanent residents include species of sea anemone and echinoderm, and a few crustaceans, the life cycles of which are adapted to the pool environment, and others that are adapted for life in the dense algal turf present in many middle- and lower-shore pools. Purely transient occupants are those that may occur in rock pools for part of each tidal cycle, or season, and include the truly accidental species, such as shoals of small, pelagic, clupeid fish (i.e. members of the family Clupeidae: herring, sprat and related species), or even the occasional, large, pelagic squid *Loligo vulgaris*, driven or attracted inshore by particular circumstances and temporarily stranded in deep, low-shore pools for the briefest period of the tidal cycle.

A rock pool may be regarded as a natural aquarium, allowing opportunities to observe and research intertidal plant and animal communities but it is also, to an extent, a natural laboratory. Each pool is unique, with physical environmental characteristics that change through tidal cycles and seasons, in relation to its position on the shore, and in its dimensions and depth. The biological community of the pool responds to these changes – its diversity, density and the patterns of growth and reproduction of its constituent species all reflecting the changing environment of the pool. Physiological characteristics and adaptations of each species drive biological responses to fluctuations in the physical environment, and measuring these responses requires specialist equipment and techniques. Yet there is still a great deal of basic ecological information to be gathered from well-planned fieldwork.

It might be noted that present understanding of the ecology of north-west European rock pools is at least partly based upon insights gained from several classic studies of rock-pool habitats on other coasts, especially those of New England (USA) and New South Wales (Australia), and that many local studies on British rocky shores have still to be tested on wider scales. Physical parameters, such as temperature and salinity, and their daily and seasonal fluctuation, in relation to tidal cycles, tidal elevation, aspect and wave exposure, can be recorded using the simplest equipment, and provide frameworks for comparison of biological communities between pools and shores. For many rock-pool invertebrates, details of life cycles are often incomplete, or described for only a narrow range of pool habitats, and for others are largely unknown. Simply recording presence, abundance and seasonal occurrence adds useful information regarding distribution patterns of even relatively well-known species, important in a time of accelerating environmental change. Regular monitoring of particular pool habitats may show significant changes in reproductive cycles and population dynamics, which are also poorly recorded for many common taxa.

2 The pool environment

Coastal topography and geomorphology are among the primary determinants of pool habitats. Steep, wave-battered coasts may have only a narrow intertidal zone, while on lower profile shores a broad extent of intertidal rock may be uncovered at tidal emersion. Soft rocks such as chalk and shale erode swiftly, and are often subject to periodic collapse, and pools created by the sea's erosive force may not persist for long. The hard igneous and metamorphic rocks – basalt and granites, schists and gneisses – erode so slowly that fissures and basins appear to show no change through time. Between these extremes, sandstones and limestones, especially where they have distinct bedding planes, and are folded, fractured and blocked, allow the development of the greatest variety of pools, and provide habitat for the most diverse rock-pool communities. The relict shorelines marking past sea level change, and large tidal ranges, are also significant factors. Thus, a wave-cut platform of folded and fractured Carboniferous limestone, such as those of the Gower Peninsula (Fig. 2.1), crossed by deeply eroded gullies, with overhangs and long passages worn along bedding

Fig. 2.1 A wave-cut platform at low tide, on the Gower coast.

planes, and deep, round pools created by physical erosion, will provide the most heterogeneous pool habitats.

A rock pool may be defined as an isolated body of water enclosed within a rock basin, replenished by a rising tide, rainfall or upwelling groundwaters (rock-pool habitats have been recognised in environments far from the sea!). In rocky intertidal environments, periodic tidal emersion and varying degrees of wave exposure create rock-pool environments that are subject to greater fluctuation in physical environmental characteristics than inshore sublittoral habitats. Deep gullies act as drainage channels for the ebbing tide and as obvious conduits for the returning flow; the deepest channels will retain water throughout the tidal cycle, but their physico-chemical profiles will remain more or less constant, unlike pools enclosed by isolated rock basins (Fig. 2.2).

On limestone shores, isolated, deep pools with an almost symmetrical round section are created by the combined effects of chemical solution and abrasion by large cobbles swirled continuously by wave action. The lower sides of such pools are smooth and polished, with no encrusting plants or animals. At some shore levels, pools offer a refuge during low tide periods, and the deepest pools at the lowest tidal level may also provide shelter from open coast wave action

Fig. 2.2 An enclosed rock basin on an exposed Gower shore. Note Coral Weed *Corallina officinalis* (pink) and the sparse Sea Lettuce *Ulva lactuca* (green) fringing the pool, and on the open rock surface, with the brown alga Serrated Wrack *Fucus serratus* and small kelp *Laminaria digitata* attached within the pool.

Fig. 2.3 Deep tide pools, at extreme low water of spring tides on the Gower coast.

for strictly subtidal species of animal and seaweed, and thus harbour the richest marine communities. At the uppermost shore levels the pool environment is usually hostile to all but a narrow selection of plants and animals specially adapted to their harsh environment, and the majority of intertidal plants and animals trapped in such a habitat will quickly die. On some shores, especially on limestone coasts, very large bodies of water may be retained close to the water's edge on the lowest tides (Fig. 2.3), cut off from direct connection to the sea by resistant rock reefs or other geomorphological features. They are, effectively, large pools, but actually represent temporarily isolated arms of the sea, the ecology of which is related to that of the sublittoral, benthic environment, rather than that of the intertidal zone.

The position of a pool relative to tidal level, and the consequent duration of the period of emersion to which it is subject, is an important factor determining the nature of the environment it creates (Fig. 2.4). Tidal cycles follow the 28-day lunar cycle, with the highest and lowest tides occurring just after the full and new moons; these are the 'spring tides', so called from an archaic word meaning 'swollen' or 'bursting'. Tidal movement is least at the 'neap tides', occurring at the first and third quarters of the moon. Standard tidal levels are thus mean low and high water marks of spring and neap tides.

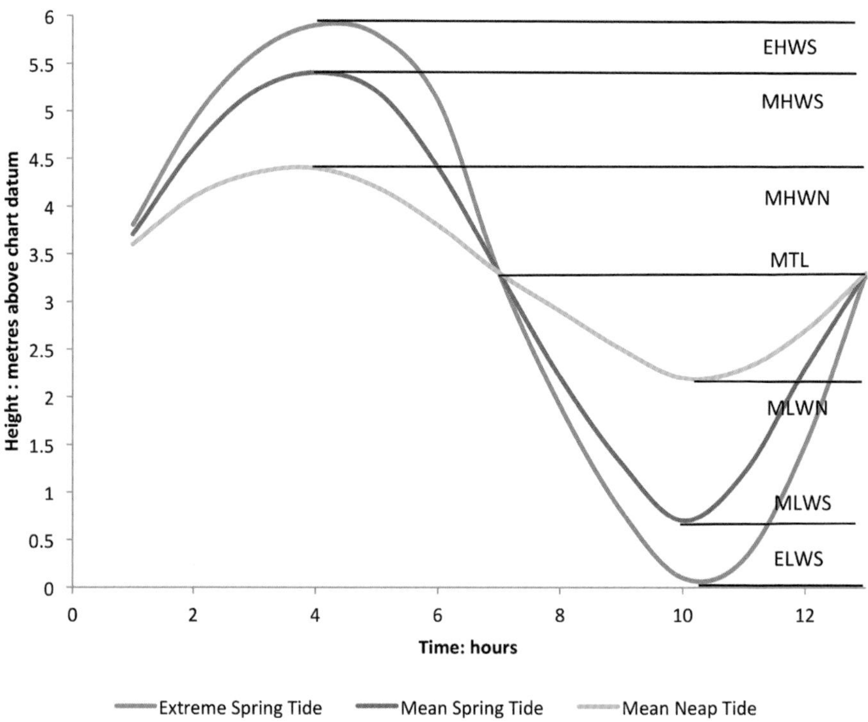

Fig. 2.4 Predicted tidal curves for an extreme spring tide, a mean spring tide and a mean neap tide, for Plymouth, showing standard tidal levels. (After Little & Kitching 1996)

Standard tidal levels are conventionally abbreviated as: extreme and mean high water of spring tides (EHWS and MHWS) and mean high water of neap tides (MHWN) above the mean tidal level (MTL); mean low water of neap tides (MLWN), and mean and extreme low water of spring tides (MLWS, ELWS) below.

Around the British Isles coastal seawater temperature shows practically no daily variation, but in rock pools temperatures may show quite sharp changes during tidal emersion, with the degree depending on the timing of the low tide period, the local tidal range and tidal cycle, and, of course, the season. The extent of temperature variation, at each of these periodicities, will depend upon the depth and volume of each pool, its tidal level and hence the length of time it is isolated from the sea – the period of emersion. Pool water salinity will also vary according to the duration of tidal emersion, the amount of sunlight and evaporation, rainfall and freshening, again modulated in relation to the depth, volume, tidal level and aspect of each pool. At the mean high water mark of neap tides (MHWN) a pool may be isolated from the sea for twelve hours, half of each daily tidal cycle, while at the mean low water mark of spring tides (MLWS) it may be cut off for less than an hour each day. Pools situated at the highest tidal level (EHWS) may be

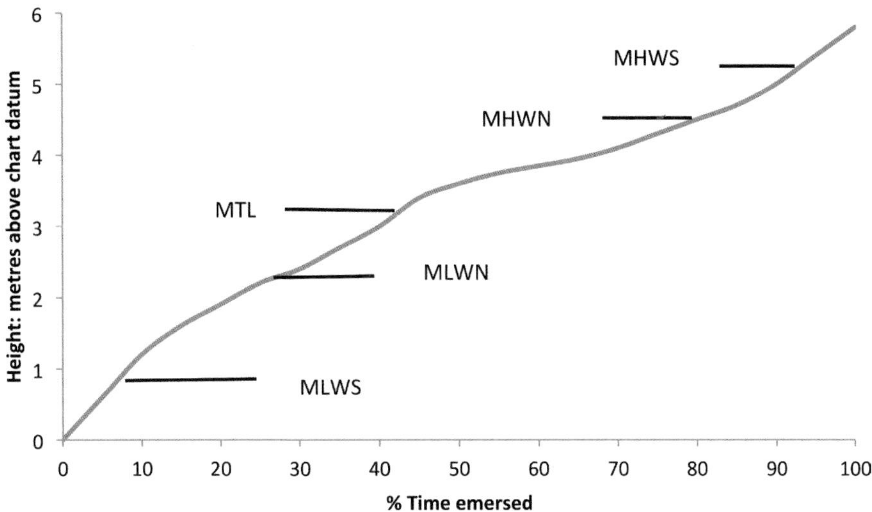

Fig. 2.5 Tidal emersion curve for Plymouth, showing the mean annual time emersed at standard tidal levels. (After Little & Kitching 1996)

flushed by the single highest spring tide and thereafter remain isolated for the rest of the fortnightly tidal cycle. Emersion curves, based on published tidal predictions, emphasise the percentage time a shore is uncovered by the sea at each standard tidal level. For example, Plymouth pools at MLWN may be emersed for almost 30% of each tidal cycle, and for close to 80% at MHWN (Fig. 2.5).

At the lowest tidal level, water temperature and salinity within the pool will vary little from ambient values for the adjacent coastal waters. Up shore, daily temperature fluctuation increases with increasing tidal level and decreasing pool depth. Temperature and salinity characteristics of each rock pool are modulated by aspect, with less impact on shaded, north-facing shores than on sunny, south-facing coasts.

In summer, the daily temperature range in shallow upper-shore pools may be greater than the annual range in local sea surface temperature, while in winter, water temperatures in high shore pools may be consistently lower. Depth and area of the pool influence daily and seasonal temperature fluctuations, and the ratio of pool area to depth is especially significant. The larger the pool, the greater the surface area for heat exchange: large, shallow pools warm and cool more rapidly than small deep pools, which may display a distinct thermal layering, with least temperature fluctuation at the bottom (Daniel & Boyden 1975*). Tidal level, and the

* references cited in the text appear in full under authors' names in 'References and further reading' on pp. 160–167

consequent duration of tidal emersion, is another factor influencing fluctuation in pool water temperatures (Fig. 2.6).

Salinity values may also show daily fluctuation greater than the annual ambient range – depressed by rainfall and freshwater runoff and rising as a result of evaporation on hot summer days (Fig. 2.7); both effects vary seasonally in relation to tidal level.

If flushed by seawater at high spring tides, high shore pools that may be essentially freshwater environments through the winter may become lethally saline as a result of rising evaporation through the summer. Sunlight, or 'irradiance', is another primary influence on rock-pool communities, displaying significant diurnal and seasonal fluctuation, in relation to day length and tidal cycle, and to less predictable factors related to the weather, including cloud cover, rainfall and resulting water turbidity. Periods of strong irradiance during tidal emersion may be particularly stressful to seaweeds, as enhanced photosynthesis results in excess production of damaging reactive by-products leading to tissue bleaching (Fig. 2.8).

Coralline algae have some defence through photo-acclimation, in which chloroplasts maximise harvesting of light during winter months, when irradiance is lowest, while displaying photo-inhibition during summer daylight emersion.

photo-inhibition
a reduction in photosynthesis that can occur at very high light levels, mainly as a result of damage to chloroplasts

Among physico-chemical factors important in rock-pool ecology, oxygen and carbon dioxide concentrations, and thus pH, are particularly significant in that they are partly determined by biological processes within the pool community. In shallow pools, warm water and sunlight promote enhanced photosynthesis by seaweed communities; oxygen concentration increases sharply, and as it exceeds saturation point free oxygen is liberated. This can be seen on sunny days in upper-shore pools occupied by tresses of the green Gutweed *Ulva intestinalis*, when bubbles of free oxygen rise to the surface. When the shore is again immersed, pool water temperatures decrease, and through the night photosynthesis is exceeded by respiration, so oxygen levels dip (Fig. 2.6).

Carbon dioxide (CO_2) dissolves readily in seawater (H_2O), forming carbonic acid (H_2CO_3), which dissociates as bicarbonate (HCO_3^-) and hydrogen (H^+) ions (Fig. 2.9A); the concentration of H^+ ions is a measure of the acidity of seawater, its pH value, and globally has a range of 8.0 to 8.5. (The pH scale ranges from 0 to 14; 7.0 indicates a neutral solution, lower values indicate increasing acidity,

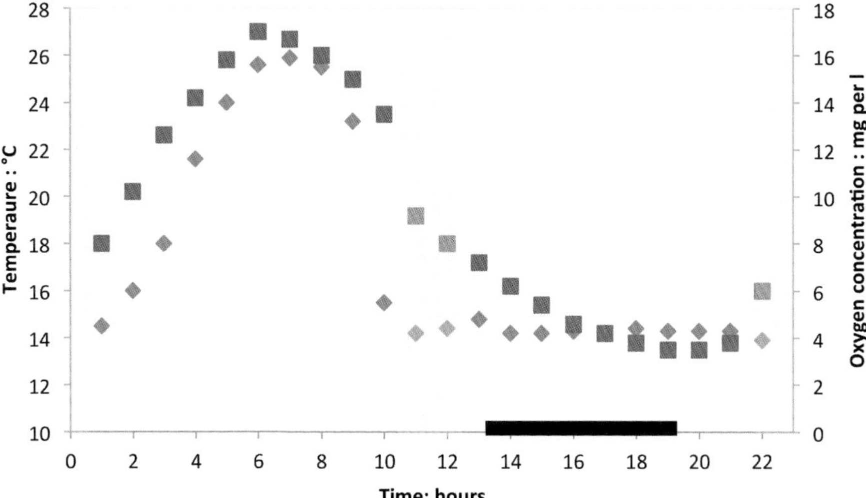

Fig. 2.6A Diurnal variation in temperature (blue) and oxygen concentration (red) in a shallow (0.14 m depth), high-shore pool (>MHWN); summer, Pembrokeshire. Shaded areas indicate immersed period; black bar represents night. (After Daniel & Boyden 1975)

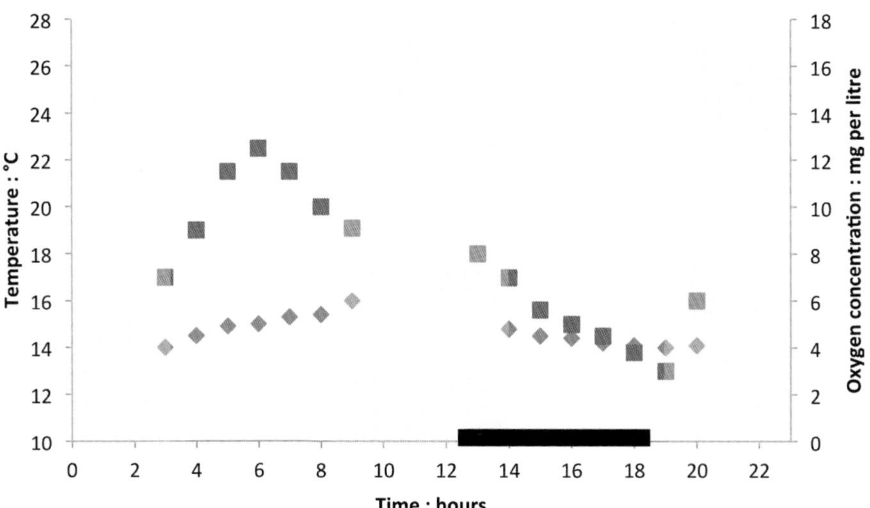

Fig. 2.6B Diurnal variation in temperature (blue) and oxygen concentration (red) in a shallow (0.26 m depth), low-shore pool (~ MLWN); summer, Pembrokeshire. Shaded areas indicate immersed period; black bar represents night. (After Daniel & Boyden 1975)

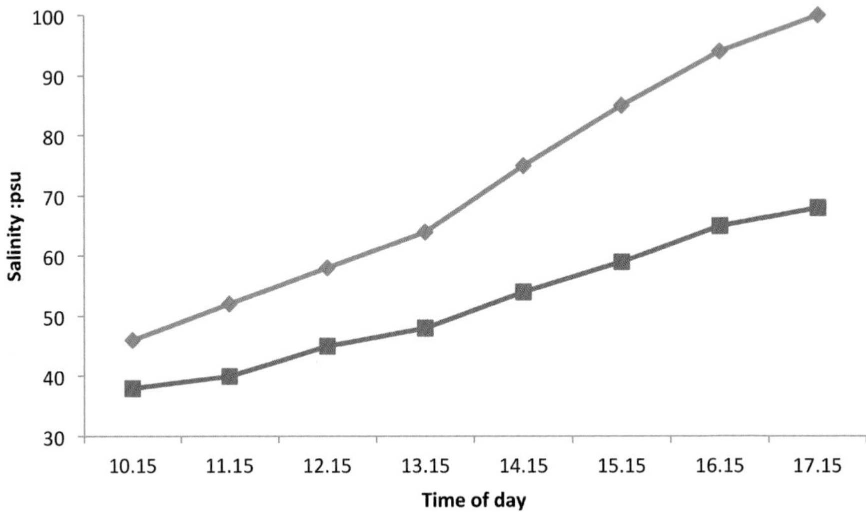

Fig. 2.7A Increase in salinity in two supralittoral rock pools at Millport, over a hot day in July. (After McAllen *et al.* 1998)

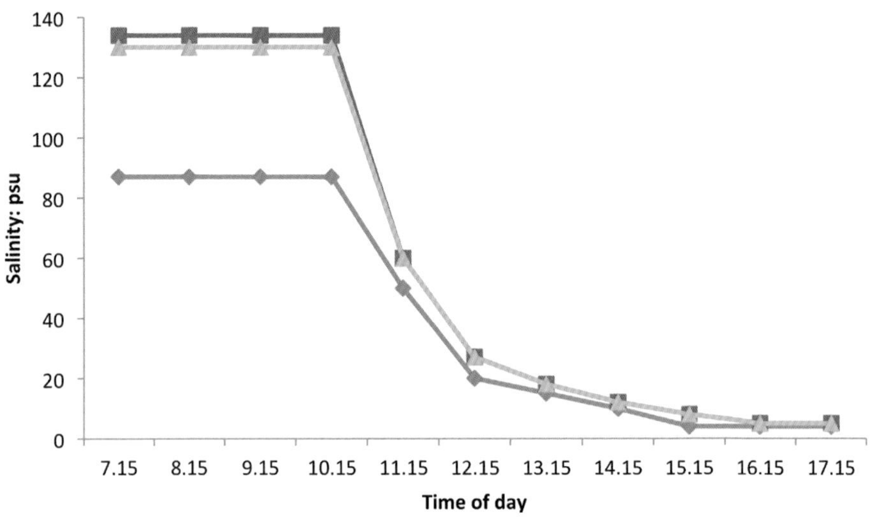

Fig. 2.7B Decreasing salinity in three supralittoral rock pools at Millport through a rainy July day. (After McAllen *et al.* 1998)

Fig. 2.8 A supralittoral rock pool in early summer: the green Gutweed *Ulva intestinalis* bleaches white as the pool dries.

higher values increasing alkalinity.) A proportion of the bicarbonate ions dissociate as carbonate ions, and further hydrogen ions (Fig. 2.9B).

Both reactions are reversible: bicarbonate and carbonate ions act as reservoirs of free carbon dioxide, as carbon dioxide is utilised in photosynthesis the reactions move to the left, ions recombine and carbon dioxide is replenished. This lowers the concentration of hydrogen ions; as a consequence, pH increases and the water becomes more alkaline. Carbon dioxide is never limited in open-sea habitats as it is continually added from the atmosphere, from the respiration of marine organisms, and through the decay or mineralisation of organic materials; hydrogen ion concentrations and pH are maintained because seawater acts as a

A. $$CO_2 + H_2O \leftrightarrow H_2CO_3 \leftrightarrow HCO_3^- + H^+$$

B. $$HCO_3^- \leftrightarrow CO_3^{2-} + H^+$$

Fig. 2.9 The carbon dioxide cycle in seawater.
A: Carbon dioxide dissolves in seawater as carbonic acid, which dissociates as bicarbonate and hydrogen ions.
B: Bicarbonate ions dissociate as carbonate and hydrogen ions.

buffered solution. However, in rock pools, during daytime summer emersion, algal photosynthesis may occur at such a rate that as oxygen levels rise, hydrogen ion concentrations drop and pH increases, to the extent that the water becomes highly alkaline. During night-time emersion the reverse occurs: oxygen decreases, carbon dioxide is released by algal respiration, H^+ increases, and pH values swing back towards the acidic end of the scale. Depending on pool volume, fluctuation in pH values may be a further source of severe stress for rock-pool communities.

Rock-pool seaweeds thus change the chemical environment of the pool through photosynthesis and respiration, but also through the exudation of metabolic by-products such as phlorotannins. These seem to have a role in interspecific competition by suppressing growth in spatial competitors, and in the case of the larger brown seaweeds may act to deter some herbivorous invertebrates. Dissolved exudates and mucus further condition the pool's aquatic environment, which in some instances becomes inimical to free-swimming invertebrate larvae. Pool communities are to some extent structured by grazers, and by interactions between predators and their prey. Limpets and winkles are significant consumers of green seaweed, especially the two species of *Ulva*, as well as sporelings and new growth of the larger brown and red algae, and are in turn preyed upon by crabs, particularly the Shore Crab *Carcinus maenas*, and fish. These are conspicuous, and their effects often apparent, but turfs of filamentous red algae may harbour large numbers of small herbivores, often referred to as 'mesoherbivores' or 'mesograzers', which may be equally important factors in rock-pool ecology.

3 Rock-pool seaweeds

The large brown seaweeds abundant on sheltered to moderately wave-exposed rocky shores display zoned distributions, with each species constituting the spatial dominant between specific tidal levels, and forming broader or narrower bands depending on the slope of the shore and its degree of exposure to wave action. The boundaries of each seaweed zone are often rather sharply marked, showing clear upper and lower limits. In general, the lower limit of each zoned seaweed species is set by competitive pressure, its downward extent limited by the next zoned spatial dominant, while upper boundaries are conventionally defined by physical environmental factors. Actually, this is just a generalisation. The upper limits of intertidal brown seaweeds may be defined by physical factors, such as temperature and duration of tidal emersion, but these factors have their primary effect on the competitive ability of the species, which depends upon its growth rate, reproductive output and recruitment. At its upper tidal limit, one species may be displaced by a more vigorous competitor, better adapted to conditions at the transition zone between the two, but it has been shown, for example, that removing plants of the high-shore alga Spiral Wrack *Fucus spiralis* will allow the lower-zoned Bladder Wrack *Fucus vesiculosus* to extend its upper limit, in the absence of its better-adapted competitor.

Zonation patterns of intertidal seaweeds are modified in response to increasing wave exposure. Bulky brown algae common on sheltered shores of north-west Europe, such as Egg Wrack *Ascophyllum nodosum*, gradually decline in abundance, and eventually disappear; two large fucoids persist, Serrated Wrack *Fucus serratus* close to ELWS and *F. vesiculosus* around MTL (mean tidal level; see Fig. 2.4), while through much of the mid-shore region broad-fronded red seaweeds, especially *Mastocarpus stellatus* (Fig. 3.1), become conspicuous.

A majority of the red seaweed species common in the rocky intertidal zone are much smaller than *M. stellatus*, forming a shrubby understorey, or filamentous epiphytic growths, although the calcified coralline algae become increasingly conspicuous on more exposed shores. Species of *Corallina*, recognised as stiff, pink clumps (Fig. 2.2), are most common, while pink to red sheet-encrusting species, belonging to several genera, become frequent on open rock surfaces. Green seaweeds tend to be small and rather

Fig. 3.1 The red seaweed *Mastocarpus stellatus* is readily recognised by its narrow frond, channelled on one side.

Fig. 3.2 The Sea Lettuce *Ulva lactuca*, a green seaweed.

inconspicuous, although a few large species are common on most shores across the wave-exposure gradient.

The Sea Lettuce *Ulva lactuca* (Fig. 3.2) grows as delicate, folded sheets, just two cell layers thick, and is often abundant in local shelter, free from grazing snails; the Gutweed *U. intestinalis* forms tangled clumps of similarly thin tubes. Neither species shows a clearly zoned distribution in the intertidal zone, but *U. intestinalis* may be common in high-shore rock pools (Fig. 3.3).

Large, deep pools and gullies also modify the distribution patterns of intertidal seaweeds. The two kelp species *Laminaria digitata* and *Laminaria hyperborea* and the Sea Oak

Fig. 3.3 High-shore pool with green tresses of *Ulva intestinalis*.

Halidrys siliquosa usually reach their upper zonal limit around the mean low water mark (MLWS) of spring tides (Fig. 2.5) but can be found in deep pools around mean tidal level on moderately exposed shores.

None of the large brown seaweeds, or the foliose reds, can be considered characteristic of rock-pool communities, although Carrageen *Chondrus crispus* (Fig. 3.4) may maintain stable populations in deeper pools on the middle and lower shore.

However, along the exposure gradient, as the zoned species diminish and coralline algae occupy more of the intertidal rock surface, enclosed pools develop a distinctive algal flora. On the most exposed shores, even the shallowest pools above MTL are colonised by sheet-encrusting corallines (Fig. 3.5). These comprise several genera, including *Lithothamnion*, and a presently uncertain number of species; identification is difficult and these taxa are often loosely termed 'lithothamnia'. In low-shore pools, these sheet-encrusting lithothamnia are favoured by settling larvae of rock limpets, *Patella vulgata* and *Patella ulyssiponensis*, and the epiphytic Blue-rayed Limpet *Patella pellucida*; the pools are important nurseries from which the juvenile limpets will disperse once they reach a size at which they are resistant to desiccation.

Fig. 3.4 The red seaweed Carrageen *Chondrus crispus*. The white encrustations are colonies of the bryozoan *Electra pilosa* (see p. 63).

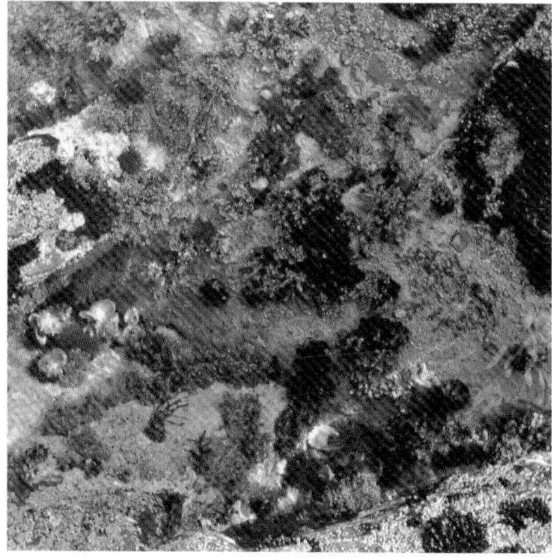

Fig. 3.5 Upper-shore rock pool on an exposed coast, with sheet-encrusting corallines, small clumps of *Corallina*, barnacles, Beadlet Anemones *Actinia equina* and patches of small mussels.

Deeper pools are usually fringed with a band of tufted, articulated *Corallina officinalis* (Fig. 3.6), extending 5 cm or more below the water surface, with populations of small, filamentous red algae, particularly *Lomentaria articulata*, species of *Ceramium* and *Gelidium*, and *Halurus flosculosus* (Fig. 3.7), forming a dense and luxuriant band below the *Corallina*. These two components of the pool flora are especially significant factors in rock pool ecology, enhancing habitat complexity and supporting diverse assemblages of small invertebrates.

The bands of *Corallina officinalis* and small red seaweeds create distinct microhabitats within the rock pool. Their dense growth, and particularly the filamentous structure of the red seaweeds, provide surfaces for the attachment of small sessile animals. They also promote the accumulation of silt and fine detritus, providing food and habitat for micro-organisms and an array of small detritivores and herbivores, and equally small predators. They support diverse communities of small arthropods, especially mites, pycnogonids (sea spiders) and crustaceans, and many species of bivalve, shelled gastropod and sea slug (see Chapter 5 for a guide enabling recognition of these groups of small animals). Some of the molluscs will be juveniles, newly settled from

Fig. 3.6 A fringe of *Corallina officinalis* at the rim of a mid-shore rock pool.

Fig. 3.7 A fringe of filamentous red seaweeds below the *Corallina* fringe, principally *Lomentaria articulata* and *Ceramium virgatum*. (Photo: www.aphotomarine.com)

the plankton, which will eventually move onto their adult substratum, usually the larger red and brown seaweeds, but for the majority of the smallest species the *Corallina* turf and the filamentous red seaweeds are their definitive habitats. The smallest arthropods – copepods, ostracods and mites – are principally detritivores or micrograzers, feeding on silt and micro-organisms, while larger animals (up to around 2 cm body length) are mostly herbivorous. This category includes small gastropods and polychaetes, as well as isopods and amphipods, and a few decapod species, which are collectively referred to as mesoherbivores or mesograzers (Brawley 1992). They are considered to have significant roles in the community ecology of their respective habitats, but such small animals are not easy to study and their impacts are often still largely a matter of conjecture. Some species seem to feed upon the seaweed they live amongst; others may inhabit crevices, live beneath stones or construct tubes within algal holdfasts, emerging periodically to feed on a wider or narrower selection of algal species. Others may feed on the micro-organisms growing upon the seaweeds and are perhaps more appropriately viewed as mesograzers. Through selective grazing, mesoherbivores might have an impact on the community dynamics of the algal fringe. Mesograzers might affect productivity or longevity of seaweeds by removing epiphytes: epiphytes might be beneficial to the host alga in reducing desiccation

holdfast
the root-like structure attaching the seaweed to its substratum

and photo-inhibition in sunlit pools, but may also lower photosynthetic rates and increase risk of loss through drag in more wave-exposed pools. Both herbivory and grazing are likely to promote algal succession and enhance species diversity by creating space for the settlement and growth of reproductive propagules of later colonising species.

succession
an ecological term describing change within a community as species replace one another over time

3.1 *Ulva intestinalis*

Intertidal species of green seaweed are mostly small, with a slender, finely branched, filamentous structure. Apart from the two species of *Ulva*, only shrubby clumps of *Cladophora rupestris* and *Codium fragile* may be at all conspicuous. The former may be found in larger pools, but more often on rock surfaces beneath the fucoid canopy, while the latter, less common species, has a sporadic distribution in low-shore pools subject to steady water flow through the tidal emersion period. The filamentous green algae are mostly small species of *Cladophora*, few of which are identifiable in the field. Most seem to occur in a wide range of intertidal habitats; they may be found in mid- and low-shore pools, below and amongst the bands of small reds, and perhaps are essentially opportunistic, attaching and growing where surfaces have been recently cleared. However, in high-shore pools, on both exposed and sheltered shores, *Ulva intestinalis* may be extremely abundant and is also ecologically important; while it does not create much in the way of habitat complexity, it is responsible for altering the physico-chemical structure of its immediate environment.

Ulva intestinalis is a fast-growing, opportunistic species, flourishing in the broadest range of coastal habitats. It will tolerate both fully saline and brackish conditions, as well as essentially fresh water, and occurs on every kind of substratum from open rock to coarse, unconsolidated gravels. Attached to the smallest pieces of shell or gravel it will even colonise sand and mud, and in particularly sheltered habitats, such as coastal lagoons, it may grow unattached to any substratum. Both *U. intestinalis* and *U. lactuca* show greatly enhanced growth and productivity in coastal habitats subject to freshwater runoff enriched by agricultural fertilisers and wastewater, or polluted by sewage effluents. The tubular thalli of *U. intestinalis* may grow to 30 cm in length and to 18 mm diameter, each attached by a small, discoid holdfast. They are mostly unbranched, with rounded tips, but both size and morphology may vary in relation to environmental factors. The intertidal distributions of both species of *Ulva* are limited by grazing molluscs, especially

thallus (pl. thalli)
the upright portion, or frond, of a seaweed

limpets (*Patella* species) and winkles (*Littorina* species), and populations are quite restricted, and often ephemeral, on most shores, including rock pools. However, *U. intestinalis* may occur abundantly in high-shore pools not habitable by grazers. It is an annual species, growing rapidly from early spring but bleaching white and dying off in the autumn. It grows as two morphologically identical (isomorphic) forms: the sporophyte and the gametophyte. Every cell of the sporophyte contains two copies of each chromosome, and it is thus referred to as the diploid form. Reproductive cells of the sporophyte undergo a reduction division (termed meiosis); the two cells resulting from the division carry just one copy of each chromosome and are termed haploid, these are then shed as microscopic, flagella-bearing spores, which settle and grow into haploid gametophytes. This phase of the reproductive cycle of *U. lactuca* is essentially asexual. However, the gametophyte then releases similar flagellated haploid cells, termed gametes, which fuse to form diploid zygotes (by definition a sexual reproductive process), that then settle, and grow to complete the cycle as a new sporophyte generation. Reproduction appears to be continuous, though periodic, through spring and summer; sporophytes usually outnumber gametophytes through much of the year, although the latter may increase to up to around 30% of the population by late summer. The overwintering form is usually a diploid zygote, but reproductive cycles vary between populations, in relation to habitat and probably to genetic differences between populations. Confusingly, the diploid sporophyte may even reproduce asexually, resulting in diploid clones. Release of the microscopic spores and gametes is synchronous, and perhaps with a tidal periodicity, and at densities sufficient to colour pool water a bright grass green.

For most seaweed species bicarbonate ions (HCO_3^-) provide the main source of inorganic carbon, as CO_2, utilised in photosynthesis. This is an evolutionary adaptation: concentrations of HCO_3^- ions in seawater are around 200 times greater than the concentration of dissolved CO_2 (Björk *et al.* 2004). There are two pathways for the uptake of bicarbonate ions. In most seaweeds they are dehydrated extracellularly, to yield hydroxyl ions (OH^-) and CO_2, the latter subsequently incorporated into the cell, but in some species bicarbonate ions are transported directly across the cell membrane and processed intracellularly, with the resulting hydroxyl ions being transported back across the membrane. Extracellular conversion of bicarbonate ions cannot occur under the

et al.
short for *et alia* = and others, used when there are a number of authors

conditions of high pH values and low CO_2 characteristic of stressed, high-shore pool habitats. However, *Ulva intestinalis* displays both facilities for HCO_3^- uptake and is able to switch to intracellular processing in response to rising pH. In enclosed pools, photosynthesis by *U. intestinalis*, utilising membrane-transported bicarbonate ions, results in reverse transport of OH^- ions and leads to a further increase in pH, to 10, or higher, while its rapid growth rate results in a depletion of organic carbon concentration. These conditions create a competitive edge for the seaweed, as algae that are dependent on extracellular conversion of HCO_3^-, such as *Chondrus crispus*, are unable to photosynthesise. On the west coast of Sweden, on a rocky shore with a tidal range of less than 0.3 m, it was shown that while the red alga *C. crispus* and the brown seaweed *F. vesiculosus* occurred together with *U. intestinalis* in shallow sublittoral bays, only the latter was found in enclosed rock pools at 0.5 m and 2 m above mean seawater level (Björk *et al.* 2004). Culturing all three species, together and singly, in the laboratory, and by transplanting plants of the red and the brown seaweeds into the pools occupied by *U. intestinalis*, suggested that photosynthetic performance in all cases was at least partly related to pH, and that significant changes in pH and CO_2 attributable to *U. intestinalis* had deleterious effects on the other two. In culture, photosynthesis by *C. crispus* resulted in pH increasing

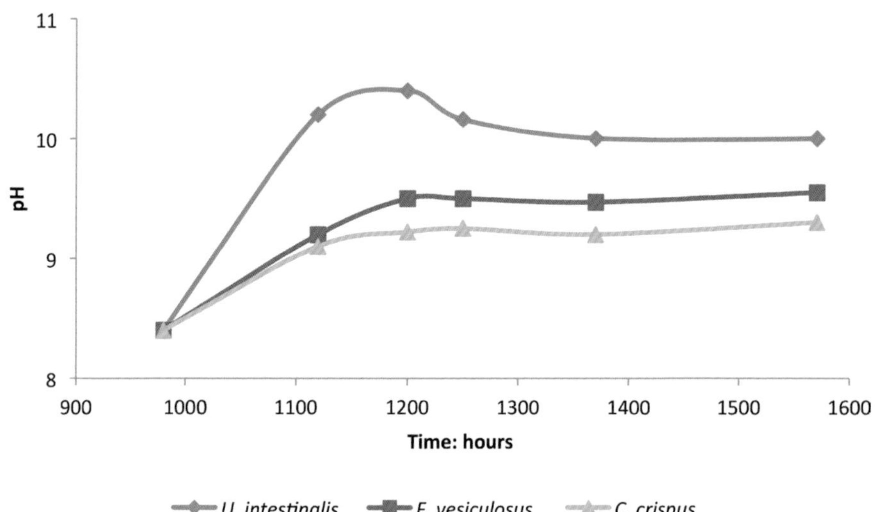

Fig. 3.8 Changes in pH values in experimental tanks for three seaweed species, cultured singly. (After Björk *et al.* 2004)

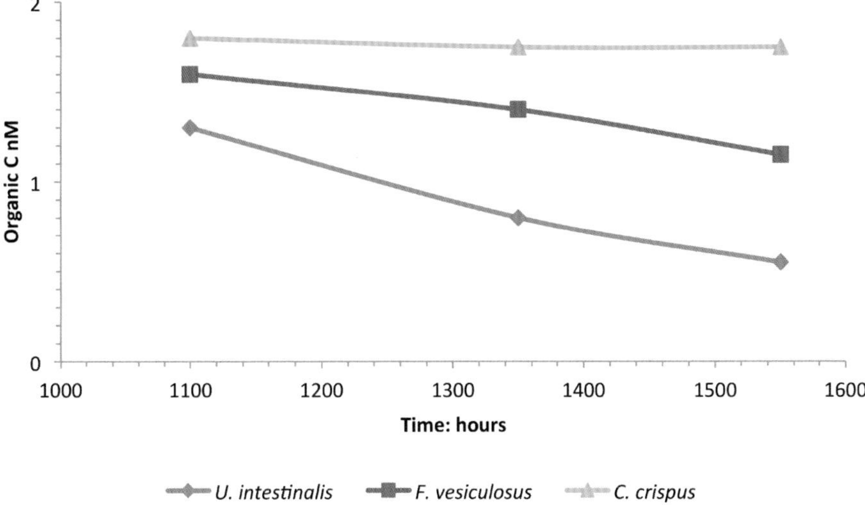

Fig. 3.9 Changes in organic carbon concentrations in experimental tanks for three seaweed species (*Ulva intestinalis*, *Fucus vesiculosus* and *Chondrus crispus*), cultured singly. (After Björk *et al.* 2004)

to 9.3, and by *F. vesiculosus* to 9.6, while *U. intestinalis* raised pH to values greater than 10 (Fig. 3.8)

In *C. crispus* photosynthetic activity ceases at pH 9.5, and in *F. vesiculosus* at 9.7–9.8. The photosynthetic activity of *U. intestinalis* in enclosed rock pools, creating pH values greater than 10, thus precludes colonisation and persistence by two potential spatial competitors. Further, rapid growth rates by *U. intestinalis* lead to a swift depletion of organic carbon concentrations (Fig. 3.9) and a further check on extracellular bicarbonate conversion.

It would be interesting to investigate the distribution and density of these three seaweed species on rocky shores around Britain in relation to different pool habitats, whether enclosed or open, and to diurnal changes in seawater pH values.

In fully saline habitats *Ulva intestinalis* rarely achieves significant density, its growth and spread being constantly checked by grazing molluscs. Its populations are essentially transient, for which reason, together with its annual lifespan, it does not sustain a characteristic assemblage of associated species. However, in high-shore pools it may provide shelter for similarly eurytopic species and it does support one animal particularly adapted to harsh upper-shore habitats: the copepod *Tigriopus brevicornis* (Section 4.1; Fig. 4.4).

eurytopic
an organism with wide environmental tolerances

3.2 *Corallina* turf

Corallina officinalis belongs to a family of red seaweeds, the Corallinaceae, within the order Corallinales (coralline algae), in which the thallus is hardened by the incorporation of calcium carbonate. The family has a practically worldwide distribution in shallow seas. Species are difficult to distinguish individually, so the taxonomy of the family is uncertain and the several dozen species presently assigned to the genus *Corallina* may in reality represent far fewer, or far more, than are currently recognised. Turfs of *C. officinalis* have been described from temperate coastal habitats in the north-east and north-west Atlantic regions, the Pacific coast of North America and Australia. Such a wide geographical distribution for a single taxon usually suggests that the name actually encompasses more than one species. Yet, molecular genetic evidence has demonstrated that populations of *C. officinalis* from north-west Europe and the north-east Pacific, from British Columbia to Alaska, appear to constitute a single genetic species (Hind *et al.* 2014). Such circumboreal distribution patterns are not unusual among marine invertebrates and algae, but distributions, especially of sedentary or sessile organisms, that extend through temperate habitats of both hemispheres are less likely, except in the case of species distributed by human agency (e.g. via shipping). The problems associated with species definitions of coralline algae based largely on morphological characteristics are highlighted by recent research demonstrating that two species of *Corallina* distributed around British and Irish coasts, and along the Atlantic and Mediterranean coasts of Europe, actually belong to distinctly separate genera (Hind & Saunders 2013).

circumboreal describes a biogeographical distribution that encompasses the entire cool temperate northern hemisphere

Corallina officinalis grows as erect, branching tufts arising from a thin but expansive, encrusting sheet. The shoots are distinctly articulated (jointed or 'geniculate' in botanical terminology), the slightly flattened articles (intergenicula) linked by narrow, uncalcified joints (genicula). They are pinnately branched, developing a two-dimensional, feather-like form. The tufts are characteristically deep pink in colour, but often lighten to pale pink or yellowish grey, and in strong sunlight bleaching to a dull white. Shoots may be up to 12 cm in length; the intergenicula of both main stems and lateral branches tend to be longer than broad, 1–2 x 0.3–1 mm, and the laterals are well spaced apart (Fig. 3.10).

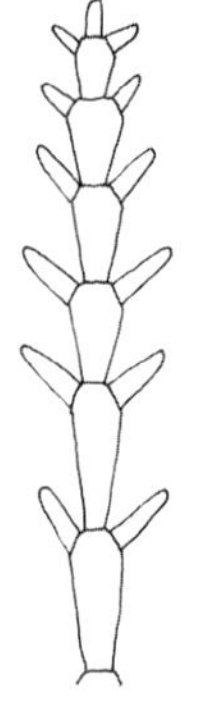

Fig. 3.10 Shoot structure of *Corallina officinalis*: apical intergenicula are longer than wide and lateral branches are well spaced.

The overall structure of the plant is influenced by environmental factors, with shorter, more densely branched and

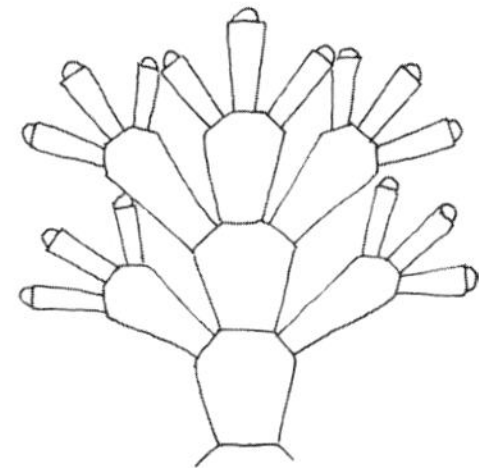

Fig. 3.11 Shoot structure of *Corallina caespitosa*: apical intergenicula with multiple lateral branches, imparting fan shape to frond.

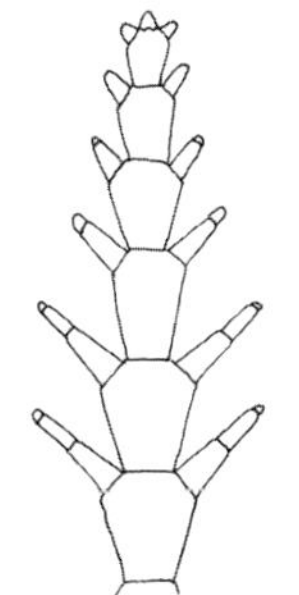

Fig. 3.12 Branch structure in *Ellisolandia elongata*: short, broad apical intergenicula with closely spaced lateral branches.

shrub-like forms on the most wave-exposed shores. It is the most commonly occurring of the erect coralline algae found around Britain and Ireland, most abundant on rocky western coasts but present wherever suitable, stable, hard substrata occur.

Two similar species of erect, branching coralline algae are presently recognised from north-west European coastal habitats (Brodie *et al.* 2013). The recently described *Corallina caespitosa* has a more southerly distribution than *C. officinalis*, recorded sporadically from south-west coasts of England and Ireland, and south to the Mediterranean (Walker *et al.* 2009). Its erect shoots are generally shorter (under 5 cm) than those of *C. officinalis* and densely branched; intergenicula are slightly longer than broad (0.6–1.2 × 0.3–0.7 mm), and often bear multiple lateral branches (Fig. 3.11).

Ellisolandia (formerly *Corallina*) *elongata* can be found in low-shore habitats on the south and west coasts of Britain and Ireland, and ranges south to the Mediterranean. Its fronds may be up to 20 cm long. Intergenicula are short, about as long as broad (0.5–1.0 × 0.4–0.8 mm), and densely branched, with successive lateral branches closely spaced (Fig. 3.12).

Distinguishing these three species correctly in the field on the basis of morphological features is difficult, especially as morphology is said to be influenced by environmental factors. It is probable that *C. caespitosa* will prove to have a wider geographical distribution than is presently recorded: densely branched plants suggestive of *E. elongata* from the Azores were genetically identical to *C. caespitosa* (Hind & Saunders 2013), and it is possible that shrubby *Corallina* from north-west European localities might be attributed to any, or all three, of these confusingly similar taxa. For the non-specialist, comparative ecological studies on *Corallina* assemblages should perhaps be related to shoot length and branch density of the population under investigation. Finally, dense tufts of a delicately structured, pink coralline alga occurring as an epiphyte of large frondose macroalgae, including species of *Cystoseira*, are referable to *Jania rubens*. Its fine structure and slender, dichotomously branching fronds distinguish it from *Corallina* and *Ellisolandia*; it occurs sporadically along south and west coasts of Britain and Ireland and is not common.

The bipartite morphology of *Corallina officinalis*, a thin, encrusting basal sheet with erect, branching fronds, is described as 'heterotrichous'. The basal sheet is sterile, and functions simply to establish and retain spatial dominance among lithophilic communities, while the erect shoots bear

lithophilic
describing organisms that encrust rock

Table 3.1 Functional-form classification of seaweeds. (After Lobban & Harrison 1997, with modifications)

Functional-form group	Morphology	Texture	Exemplars
Sheet	Thin, tubular or sheet-like	Soft, membranous	*Ulva* species
Filamentous	Finely branched	Soft	*Ceramium, Lomentaria*
Coarsely branched	Stiffly branching	Firm, wiry	*Osmundea, Gelidium*
Thick, frondose	Thick branches, stiff holdfasts	Leathery	*Chondrus, Mastocarpus*
Calcified, branching	Upright jointed	Stony	*Corallina* species, *Jania*
Calcified, encrusting	Encrusting sheets	Stony	*Corallina* species, *Lithothamnion*

the reproductive structures. Heterotrichy is an important adaptation that conveys significant competitive superiority to the alga (Littler & Kauker 1984). Each of the two morphologies represents a specific category within the 'functional-form' classification (Table 3.1) recognised by seaweed ecologists (Lobban & Harrison 1997), and their expression within a single species reflects an important evolutionary adaptation.

The crustose basal portion of *C. officinalis* is thin, tough and highly resistant to environmental disturbance; it has a low surface-to-volume ratio and correspondingly low productivity. Upright fronds, with a greater surface-to-volume ratio and thus greater access to light and nutrients, have significantly higher rates of production than the encrusting base, but are vulnerable to damage and loss through wave surge and sand scour, and to desiccation during tidal emersion. They are also more vulnerable to herbivores than the crustose base, which probably benefits from small molluscan grazers that remove the microbial films, microalgae (such as sessile diatoms) and organic detritus that would otherwise reduce its photosynthetic activity. The basal portion of *C. officinalis* has an especially important function in promoting swift recovery of the plant following severe disturbance. On San Clemente Island, southern California (USA), experimental plots within a *Corallina* turf scraped clear of upright shoots showed new growth within weeks; by six months after the perturbation they had achieved 10% of the original cover, and had exceeded 15% within 12 months. Conversely, experimental plots that were scraped and then blowtorched to remove both crust and shoots showed scarcely 10% recovery of the original cover within a year (Littler & Kauker 1984). The basal portion of the *C. officinalis* plant is thus the source of its resilience and competitive strength in challenging environmental conditions, but the equal significance of the erect portion of the plant is demonstrated by the observation that expansion of the crust is unable to continue until regeneration of erect shoots has begun, providing the energy necessary for growth of the entire plant.

Corallina officinalis thrives on open rocky coasts under most degrees of wave exposure. On sheltered, weedy shores dominated by large brown algae it will be found as small, discrete patches from the mid-tidal level seawards, and will be present offshore to about 20 m depth. It will be largely absent from very sheltered shores blanketed by tangled ropes of the Egg Wrack *Ascophyllum nodosum*, but as the bands of brown algae thin out with increasing wave exposure, *C. officinalis* becomes abundant. It grows as thick fringes around the rims of rock pools, spreading out onto wave-washed open rock surfaces, and ranging upshore in high-level pools. The pool environment changes in relation to both wave-exposure and tidal-emersion gradients, and while pools provide a refuge from dehydration during emersion, and from wave scour during immersion, their communities are subject to potentially damaging environmental stresses as a consequence of fluctuating physical factors.

Corallina officinalis loses colour when stressed, fading from the familiar deep pink to a pale yellowish grey, and eventually bleaching to a dull off-white. Prolonged bleaching results in death, and in small, high-shore pools, during warm summer weather, *Corallina* plants are reduced to chalky, skeletal white clumps (Fig. 3.13). This loss of colour results from the degradation of photosynthetic accessory pigments (phycobiliproteins) associated with the chloroplasts of the alga, and functioning in the electron transfer processes by

Fig. 3.13 Bleached clumps of *Corallina officinalis*.

which captured light energy is passed to chlorophyll *a*, the most important photosynthetic pigment.

Sunlight and temperature are most important as causes of stress-induced bleaching in *Corallina* and *Ellisolandia*, as both are positively related to photosynthetic rates. Irradiance (the amount of energy received per unit area) varies seasonally in relation to day length, predictably in relation to tidal cycle and unpredictably in relation to weather. Its greatest impact is registered in shallow high-shore pools, with the longest period of tidal emersion, during clear summer weather. Temperature shows seasonal variation and also varies through daily periods of tidal emersion, and in relation to pool size and tidal level. The mechanisms behind bleaching are not completely understood, but when related to rising water temperature appear to indicate the overloading and eventual failure of cellular defence mechanisms (Latham 2008). Hydrogen peroxide (H_2O_2) is produced as a by-product of photosynthesis. It is a powerful and damaging oxidising agent, which is eliminated from the algal cells by a sequence of enzyme-catalysed reactions: the enzyme bromoperoxidase is responsible for the uptake of bromine anions (Br^-) from seawater which react with hydrogen peroxide to form either hypobromous acid (HBrO) or bromine cations (Br^+); these are then neutralised by oxidation of organic molecules to produce volatile compounds. Some minor bleaching of pigments may be caused by the hypobromous acid or by reactive oxygen molecules released by the first reaction. However, experimental evidence suggests that high temperatures lead to rising concentrations of H_2O_2, increasing enzyme activity and continued production of both oxidising hypobromous acid and bromine cations (Latham 2008). The pool of organic molecules in the algal cells is eventually exhausted and pigment bleaching ensues. Enhanced bromoperoxidase activity has also been recorded in *C. officinalis* subjected to low salinity and high ultraviolet light levels (Latham 2008), suggesting bleaching responses similar to those induced by temperature fluctuations.

ions with a negative charge are termed **anions**, those with a positive charge are **cations**

A field study of the two *Corallina* species and *E. elongata* in rock pools on the north and south coasts of Devon found that all three showed long-term (i.e. seasonal) and short-term (hours) acclimation to changing light intensities through low-tide periods (Williamson *et al.* 2014). The plants' photosynthetic performances were recorded in relation to sunlight and temperature at the beginning, middle and end of daylight low tides, in both summer and winter. Irradiance was significantly greater during summer low tides in north coast upper-shore pools, although water temperatures showed no difference in either summer or winter. On the south coast, significant differences in irradiance were recorded through low-tide periods in both summer and winter for upper-shore pools, but not for the lower shore, while temperature increased significantly at both levels. In all three species photosynthetic rates were significantly reduced during summer tides, and negatively correlated with sunshine and temperature, but enhanced during winter low tides, when sunlight was reduced.

The density and diversity of animal communities associated with coralline algal turfs depend largely upon the structure of the plants, which in turn is determined by environmental processes. In rock pools on wave-exposed shores, the *C. officinalis* fringe consists of short-fronded, richly branched plants densely packed around pool rims, with progressively longer-fronded, less-branched plants at successive depths within the pool. It is, of course, possible that this observation, frequently made, may reflect different ecological tolerances of two or more similar species. The character of the associated fauna may be directly related to these changing morphologies, as dense turfs provide more secure refuges than open-structured plants, or indirectly so, in that they accumulate and retain greater volumes of silt, providing microhabitats for a variety of meiofaunal animals, as well as potential food sources for the macrofauna. The relative rigidity of coralline algal fronds also allows secure attachment for numerous sessile or encrusting animals, including a common, but inconspicuous, bryozoan *Walkeria uva* (Chapter 5, B.23) and the tubeworm *Spirorbis corallinae* (Chapter 5, J.6) which, as its name suggests, is an obligate epiphyte of *C. officinalis*. The invertebrate fauna of *C. officinalis* turfs is often very rich, as evidenced by a study on two shores, with similar exposure ratings, on the coast of Anglesey (Bussell *et al.* 2007). To investigate the fauna in relation to depth within pools, 25 cm^2 core samples were collected at the surface and at three successive 10 cm depth intervals from five large (>5 m × 1 m) mid-shore pools at each site, yielding a total of 10,371 individual organisms of 91 taxa. There were significant differences between the communities recorded at each of the two shores, but annelids were the dominant in both cases, followed by crustaceans, molluscs and pycnogonids. Small bivalves (*Lasaea adansoni* and juvenile *Mytilus edulis*), small snails (*Rissoa parva* and *Onoba semicostata*) and *S. corallinae* showed different abundances between the two shores; *S. corallinae* and juvenile *M. edulis* accounted for 20% of the dissimilarity between the two shores and 20% dissimilarity

meiofauna
literally, 'smaller' animals, defined as those that will pass through a sieve mesh of 1.0 mm but are retained by a mesh of 0.05 mm; some ecologists recognise an upper limit of 0.5 mm and the standard 0.063 mm sieve mesh as the lower. Larger are 'macrofauna', smaller 'microfauna'

between the shallowest and deepest depths within the pools. Both the diversity and composition of the fauna changed with depth within the pools, as did habitat complexity. Diversity was greatest in the dense, silt-rich turf at the pool rim, and declined among the more sparsely branched, long-fronded plants deeper within the pool. Encrusting species such as *S. corallinae*, presumably vulnerable to wave shock, were most abundant deeper within pools, while those living within the turf, such as the small molluscs and the brittle star *Amphipholis squamata*, were most abundant at the shallowest depths. Samples were also taken at five small (<1 m) and five large pools randomly selected from mid- and lower-shore levels to examine the faunal assemblages in relation to pool size and tidal level; a total of 14,206 individuals for 101 taxa was collected. The number of species (i.e. species richness) recorded varied significantly between shores, tidal height and pool size; for example, small, low-shore pools at site 1 and small high-shore pools at site 2 both showed greater species richness than small, low-shore pools at site 2.

3.3 The red algal fringe

Fine-structured, filamentous red seaweeds are among the most difficult algae to identify with confidence. In many species, features such as frond length, and mode and frequency of branching vary in relation to environmental stresses, while reproductive cycles are particularly complex, involving two or three stages, which may be isomorphic or heteromorphic (morphologically dissimilar). Accurate identification may depend on structures visible only with the aid of a hand lens, or even require microscopical examination. Large, foliose red algae, such as *Rhodymenia pseudopalmata* and *Delesseria sanguinea*, may be sporadically distributed in deep, low-shore pools. Small, filamentous species typically occur as a discrete band below the *Corallina* fringe, with species involved varying in density and relative abundance according to tidal level and wave exposure regime of each pool. While reproductive modes, life cycles and the morphology of each life-cycle phase may be known, for many species there is often little information relating to growth rates, longevity and population dynamics. Most red algae are dioecious (i.e. have separate male and female plants) but only the male sheds gametes into the water, those of the female remaining attached to the parent. Further, unlike the fucoid seaweeds and *Ulva*, the male gametes (spermatia) are non-flagellated. Both fucoids and *Ulva* limit loss of gametes by releasing them during periods of calm weather or during low tide, although the cues stimulating release are not clearly understood. For small red algae, the potential for loss of passively dispersing male gametes must be considerable; it is not evident that this can be limited, but it seems possible that successful fertilisation is aided by the rock-pool environment. Unfertilised female plants of *Gracilaria gracilis* (a slender, stringy species occurring sporadically on south and west coasts of Britain), transplanted into upper-shore and low-shore pools during both high-tide and low-tide periods, showed the greatest density of fertilised zygotes on plants located in upper-shore pools during low tide. However, this result is not easily explained (Engel & Destombe 2002). A longer period of tidal emersion in upper-shore pools might result in a greater number of fertilisation events, promote a greater release of male gametes or simply reflect a lower rate of dispersal and a greater concentration of gametes. It is not known whether the release of spermatia by *G. gracilis* is cued by environmental factors, and thus whether it is synchronous or continuous, or if the shorter emersion periods affect their release in low-shore pools.

A common component of the red algal band is *Lomentaria articulata* (Fig. 3.14) although it may be as abundant on shaded, damp rock surfaces on the lower shore as in pools. It is readily recognised by its long (to 10 cm), slender, deep red fronds, each consisting of shiny, hollow, cylindrical segments (internodes) up to 10 mm long and 5 mm diameter, linked by constricted joints (nodes) at which additional cylinders branch, dichotomously or in whorls. Distinctively, the tip of each frond bears a tiny pair of developing internodes,

Fig. 3.14 Structure of the red alga *Lomentaria articulata*.

resembling rabbit ears. *Lomentaria articulata* grows in dense clumps, attached by small, disc-like holdfasts, often in association with species of *Ceramium*. About 14 species assigned to this latter genus occur on north-west European coasts; they are all very similar and practically impossible to distinguish individually in the field. *Ceramium virgatum* (Fig. 3.15) is the species most common in intertidal rock pools, but it can also be found as an epiphyte of larger algae (including *Mastocarpus stellatus*), and on shaded rock surfaces, on the lower shore. It may grow to 30 cm in height, but its slender, tapered fronds are usually no more than 1 mm in diameter, translucent and cross-banded, and with pincer-like (forcipulate) tips.

Fig. 3.15 Frond structure of the filamentous alga *Ceramium virgatum*.

Fig. 3.16 Frond structure of the small red alga *Gelidium spinosum*.

Halurus flosculosus grows as clumps of stiff, wiry shoots, up to 20 cm long but even more slender than *C. virgatum* (around 0.5 mm diameter), each consisting of a single chain of cells. Two species of *Gelidium* are also typical of coralline rock pools; *G. spinosum* (Fig. 3.16) is an epiphyte of *C. officinalis*, and is recognised by its flattened, irregularly branched fronds, bearing small, lateral offshoots, also in irregular arrangement. *Gelidium pulchellum* differs from *G. spinosum* in its round-sectioned main branches (although lateral offshoots may be tapered and flattened), and in its greater tolerance of light, occurring in well-lit pools, often associated with *C. officinalis*, and on the shells of large, old limpets.

Filamentous red algae are more efficient sediment traps than *C. officinalis*, accumulating greater quantities of silt per unit of seaweed than the stiff fronds of the latter. Similarly, they are more effective in filtering and retaining invertebrate larvae. It was considered that the high densities of small gastropods typically recorded in samples of finely structured red seaweeds was a consequence of substratum selection by their settling larvae (Fretter & Manly 1977). However, noting that the sizes of several gastropod larvae were within the size range of silt particles accumulated by particular algae species, Wigham (1975) suggested that rather than reflecting choice of a preferred host seaweed, the accumulation of larvae on fine-fronded algae was simply a passive process. This was demonstrated by the positive relationship between silt load and larval density for two filamentous reds.

The variety of larval forms trapped by red algal sieves does not necessarily correlate with the composition of the resident fauna. Following settlement, newly metamorphosed juveniles may subsequently migrate to a preferred microhabitat. Most filamentous red algae tend to be delicate, flexible plants; some species may be perennial, whereas in others all life-cycle phases may be comparatively short lived. They do not provide suitable attachment sites for sessile invertebrates, and most species living within the red algal band tend to be small, mobile animals, such as the smaller snails, sea slugs and isopods (which are mostly herbivorous) plus small polychaetes, small crustaceans (such as detritus-feeding copepods and ostracods) and predatory sea spiders. Some data from a field study of three small gastropod species associated with seaweeds at Wembury, South Devon, illustrate these points neatly (Fig. 3.17). *Lomentaria articulata*, the most commonly occurring component of the red algal band, supported the greatest density of each of three snail species. However, the most abundant of these, *Lacuna vincta*,

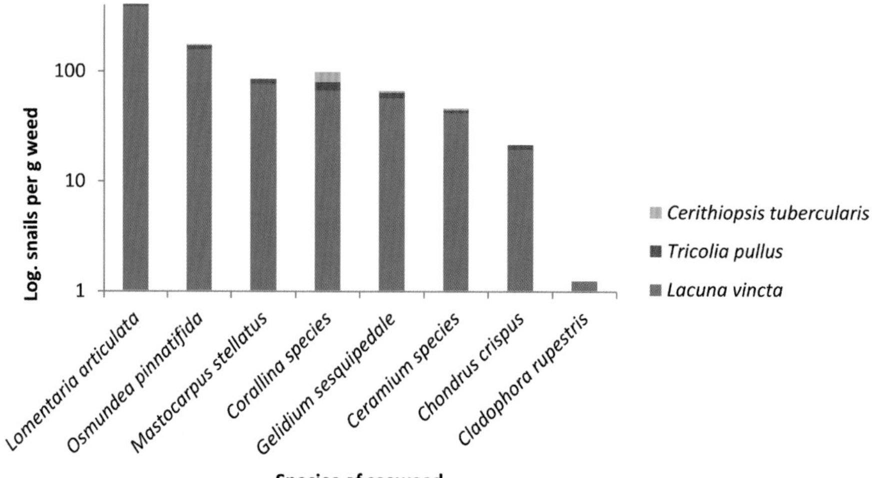

Fig. 3.17 Mean occurrence of three snail species on intertidal seaweed species at Wembury, Devon, June to September. Log numbers of snails per gram dry weight of alga. (Data from Fretter & Manly 1977)

is an epiphyte of and feeds on large fucoid algae, and it must be assumed that the algae simply served to sieve their larvae from the plankton, and that post-metamorphosis the juvenile snails would migrate to their usual host plants. *Tricolia pullus* is a grazer, feeding on sessile diatoms, as well as on algal tissue, and has no preferred host alga. Following metamorphosis, *Cerithiopsis tubercularis* would move onto the sponges growing among the algae, on which the adult snails feed. High densities of *Lacuna vincta* were also recorded on *Osmundea pinnatifida*, whereas on *Mastocarpus stellatus* its numbers were scarcely more than those on *Corallina*, *Gelidium* and *Ceramium* species. *Osmundea pinnatifida* and *M. stellatus* are coarse-fronded seaweeds: neither forms part of the red algal band, *O. pinnatifida* typically occurring as dense growths on open rock surfaces, and it is possible that their *L. vincta* populations originated as larvae settled elsewhere.

4 Rock-pool animals

primary production the process by which photosynthesising organisms utilise energy derived from sunlight to create organic molecules, such as simple carbohydrates, from water and carbon dioxide; in the sea the most important primary producers are the phytoplankton

The density and diversity of intertidal, rocky-shore animal communities fluctuate seasonally. Most motile intertidal invertebrate and fish species are seasonal migrants, moving inshore in spring in response to rising sea surface temperatures and increasing primary production, then moving offshore again as temperatures decline in the autumn. Migrations are typically pulsed, with different species of fish and decapod arriving in succession from early March into the summer. The Shore Crab *Carcinus maenas*, Velvet Swimming Crab *Necora puber*, Edible Crab *Cancer pagurus*, Common Lobster *Homarus gammarus*, plus the prawns *Palaemon elegans* and *Palaemon serratus*, are the most commonly seen decapods. Numerous, smaller species of squat lobster, crabs and prawns also participate in this spring onshore movement. The most frequently encountered fish are the Common Blenny *Lipophrys pholis* and the Butterfish *Pholis gunnellus*, but many other small species, including the pipefish *Nerophis lumbriciformis* and several species of goby, also spend summer between the tides, and some establish individual territories which they maintain through their breeding period. The permanent motile residents of intertidal rocky shores, such as winkles, topshells and dog whelks, may move into and out of pools in order to feed or to seek refuge, and their movements may reflect diurnal, tidal and seasonal cycles.

While numbers of species and individuals of animals found in rock pools vary seasonally, and in relation to tidal level and wave exposure, there are only few species that could be considered as especially adapted to the rigours of the rock-pool environment. Sea anemones, though not entirely sedentary, generally move infrequently, and may form permanent colonies in pools with suitable physical environmental conditions. Populations of the Dahlia Anemone *Urticina felina* (Fig. 4.1), for example, will persist in deep, shaded pools as far upshore as mean high water mark, provided that temperature and salinity fluctuate minimally through the tidal cycle.

The much smaller *Sagartia elegans* (Fig. 4.2), which occurs as five distinct colour variants, forms small clonal groups in shallow mid-tidal pools on moderately exposed shores, and the Snakelocks Anemone *Anemonia viridis* (Fig. 4.3) can be found as extensive clonal populations in sunlit low-shore pools.

Fig. 4.1 A group of Dahlia Anemones *Urticina felina*. (Photo: Joanne Porter)

However, a small number of macroinvertebrate species are particularly adapted to life in intertidal pools. The tiny copepod *Tigriopus brevicornis* (Fig. 4.4) – orange coloured with a maximum length of 1 mm – lives in supralittoral pools above MHWS (see Fig. 2.5) where it experiences salinities above 100 psu and temperatures that may vary widely and sharply each day. Two small echinoderms, the brittle star *Amphipholis squamata* and the cushion star *Asterina phylactica,* occur in mid- to upper-shore pools on wave-exposed coasts; their life cycles, involving hermaphroditism and brooded

psu
practical salinity units: a measure of salt concentration based on seawater conductivity, equivalent to g/kg and often expressed as parts per thousand or ppt

Fig. 4.2 The small anemone *Sagartia elegans* var. *rosea*.

Fig. 4.3 The distinctive tentacles of two Snakelocks Anemones *Anemonia viridis*, in a crevice within a small, exposed shore pool.

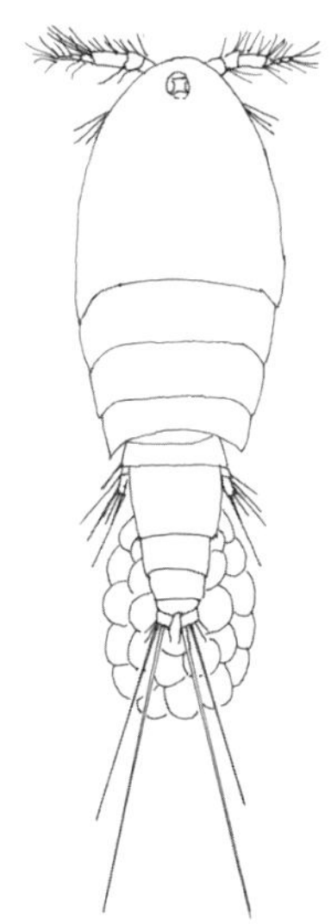

Fig. 4.4 The harpacticoid copepod *Tigriopus brevicornis*: an adult female carrying an egg mass.

larval development, are particularly adapted to their habitats. Additionally, sessile or motile organisms restricted to microhabitats within *Corallina* turf or filamentous red seaweed are essentially permanent residents of the rock-pool fauna.

4.1 Life on the edge

fusiform
spindle-shaped

Harpacticoid copepods are small, mostly free-living crustaceans with slender, fusiform bodies; few species exceed 2.5 mm in length. They are benthic (bottom dwelling) for the most part and occur commonly in intertidal habitats, burrowed into soft sediments (i.e. are infaunal) or in silty habitats beneath stones and amongst finely structured seaweeds. *Tigriopus brevicornis* (Fig. 4.4) occupies the harshest maritime habitat. It is confined to supralittoral pools that might be flooded by seawater on the most extreme high tides, but which are otherwise replenished irregularly only by wave splash or rainfall. It flourishes in even the smallest pools, frequently just a few mm deep with diameters of a few centimetres. In such pools, physical environmental conditions show the most extreme variation: salinity and temperature rise or fall on hourly and daily bases, the shallowest pools practically dry out during hot summer periods, on cloudless days are subject to damagingly high levels of ultraviolet radiation, and may freeze completely during the coldest winter. Wave surge may pose a significant physical risk, dislodging and dispersing individuals. Unsurprisingly, *T. brevicornis* displays morphological and physiological adaptations that allow it to persist in these marginal habitats.

In contrast to soft-sediment infaunal harpacticoids, species of *Tigriopus* are furnished with long, curved spines at the tips of their appendages which enable them to retain a firm grip on the microtopography of the pool floor and thus to resist dislodgment by wave action (McAllen 2001). *Tigriopus brevicornis* will survive the drying out of its habitat for days, or perhaps weeks, by assuming a dormant state; as with most harpacticoids, to avoid desiccation it will burrow into such sediment as its pool accumulates and retains, and, when present, will also take refuge beneath and within the hollow, tubular thalli of *Ulva intestinalis* (McAllen 1999). All species of *Tigriopus* will tolerate the broadest range of temperature and salinity, and the limits to both are to an extent interdependent. *Tigriopus brevicornis*, acclimated to a temperature of 10°C in water at 34 psu, was found to have a median upper lethal temperature limit of 34.9°C and a lower median lethal limit of –13°C. At 10°C the median upper lethal limit for salinity was 161 psu and the lower just 7 psu; for animals acclimated to 21°C, lethal limits were 94 psu and 7 psu respectively (Davenport *et al.* 1997). It is defined as a euryhaline osmoconformer: the osmotic concentration of its body fluids reflects that of its environment across an unusually broad range of salinities (McAllen *et al.* 1998). As pool salinity increases, the body length and volume of the copepod also increase, as does the concentration of sodium ions in its body fluids. A consequence of this is that as pool water density rises with increasing sodium ion concentrations, so does that of the animal's body fluids, ensuring that it remains negatively buoyant (Fig. 4.5). This is particularly important as *T. brevicornis* is a bottom feeder, grazing microbial films, sessile diatoms and encrusting microalgae, and browsing on the fronds of *U. intestinalis*. By contrast, nauplius larvae of the osmoregulating brine shrimp *Artemia* sp., showed minimal variation in the body fluid concentration of sodium ions over a wide salinity range.

In shallow supralittoral pools, where cover is often limited, sunny days pose a risk of damage by ultraviolet radiation. The bright orange coloration of *T. brevicornis* is imparted by a carotenoid compound, astaxanthin, and carotenoids

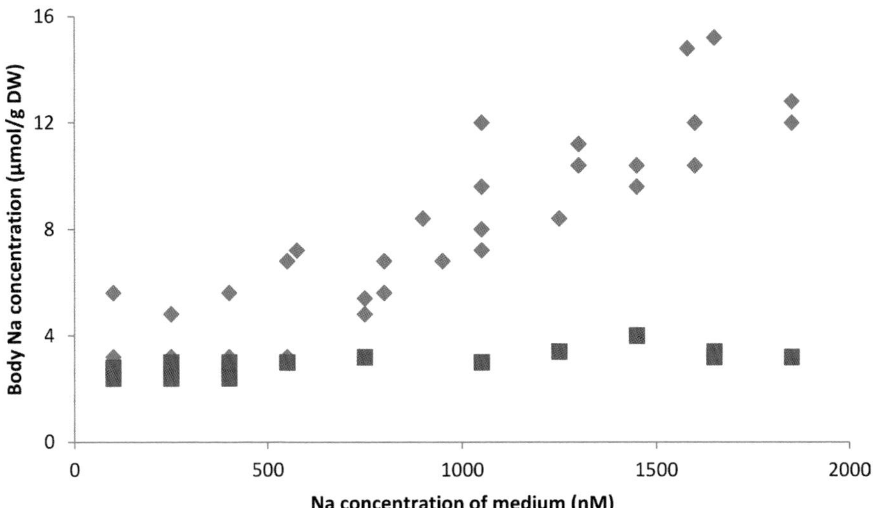

Fig. 4.5 The relationship between body sodium concentration in *Tigriopus brevicornis* (blue) and the nauplii of a brine shrimp *Artemia* sp. (red), as µmol sodium (Na) ions per mg dry body weight (DW), and salinity of the medium (nM sodium). (After McAllen *et al.* 1998)

are known to be effective shields against UV radiation. The copepod is unable to synthesise the compound itself but instead derives it from its diet, diatoms and, especially, *Ulva intestinalis*. The effectiveness of its carotenoid shield was demonstrated experimentally by exposing animals reared on a diet of dried bakers' yeast, and thus lacking astaxanthin, to UV radiation levels comparable to those recorded at noon on both cloudy and cloudless days (Davenport *et al.* 2004). The median lethal time for these animals was 21.1 hours, while wild-collected specimens survived indefinitely, beyond 96 hours. All littoral algae accumulate a range of secondary metabolic compounds in response to UV exposure, which then act as direct UV screens or antioxidant agents, limiting or preventing damage to the plant, and it is likely that a number of these will be sequestered by herbivorous animals (Ó Corcora *et al.* 2016). Experiments showed that *T. brevicornis* fed on bakers' yeast over a period of weeks became increasingly sensitive to UV, and displayed a sharp reduction in the median lethal exposure time (Fig. 4.6). When their diet was switched to *U. intestinalis* cultured under a UV light source, the median lethal exposure time returned close to its initial value, while those fed algae grown under a UV shield showed only slight recovery. *T. brevicornis* is likely to ingest a range of different secondary metabolites

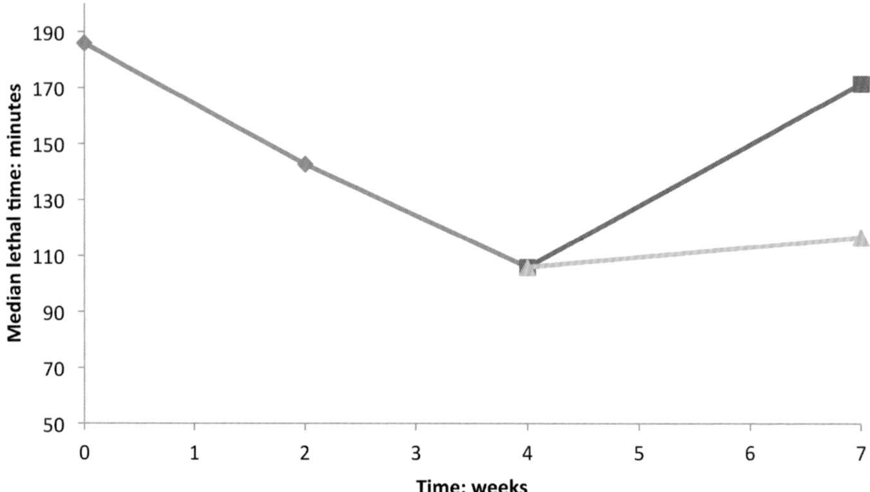

Fig. 4.6 Median lethal time for wild-collected (at time 0) *T. brevicornis* exposed to high-intensity UV radiation: fed for 2–4 weeks on bakers' yeast (blue) and then fed on UV-acclimated (red) or non-UV-acclimated (green) *U. intestinalis*. (After Ó Corcora *et al.* 2016)

from its various food sources; of these, it has been shown to accumulate the carotenoid astaxanthin, and while this seems to be the probable source of its UV protection, it remains to be proved.

4.2 Within the fringe

In discrete, clearly bordered pools on the middle and lower littoral zones of moderately wave-exposed shores, the fringe of *Corallina* turf and filamentous red seaweeds will support the greatest diversity and density of invertebrate animals. They are all small to very small animals; the greatest proportion of the total biomass of a pool fauna may consist of a few species of large gastropods, such as winkles and topshells, but species richness and total number of individuals will be greatest in the algal fringe. This algal-associated community will vary in composition, diversity and density in relation to tidal level and degree of wave exposure, which ultimately determine the complexity of the habitat: densely branched, closely packed turfs support the richest faunas and lax, diffuse growths the poorest. On moderately exposed shores, molluscs, crustaceans, polychaetes and a few species of echinoderm comprise the principal components of the algal macrofauna. Their relative abundance, and the relative abundance of each species, will differ between shores, and between tidal levels, and will vary seasonally, but they will still comprise the greater part of the algal macrofauna. The meiofauna is also likely to contribute significantly to the total species richness of algal turf communities but is not as well documented as the macrofauna.

The biology and ecology of some algal-associated macrofaunal species are reasonably well known, while for others little has been documented. Easily identified species, including many shelled gastropods, some crustaceans and the three echinoderms characteristic of pool habitats, have attracted most research effort, but for others there is very little information, especially among the less readily distinguished groups. The polychaete fraction consists of numerous individually small species, usually representing few families, among which the Syllidae are often predominant, and which are often difficult to assign even to genus. Consequently, ecological and biological studies on these demand specialist knowledge, and considerable time and effort. Similar difficulties arise with the meiofauna, on which, again, there has been only limited research focus.

4.3 Gastropods

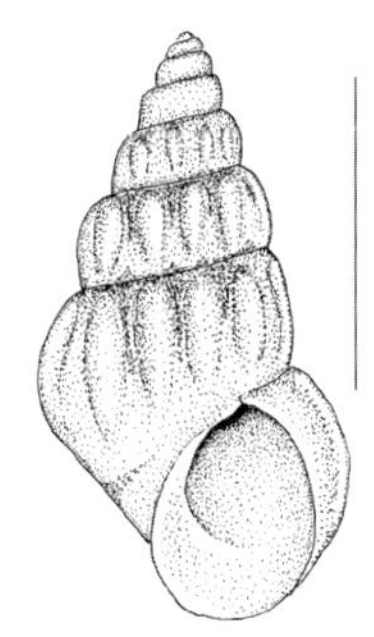

Fig. 4.7 The shell of *Rissoa parva*. Scale bar: 2 mm.

Small shelled gastropods are common and consistent occupants of the algal fringe. They are also taxonomically very diverse, with perhaps 10 or more families to be found in rock-pool habitats around the British Isles. Predominant among these are the rissoids – family Rissoidae – 20 species of which are recorded for shallow coastal habitats, including five likely to occur amongst filamentous red seaweeds. None of these small snails exceeds a shell length of 5 mm; *Rissoa parva* (Fig. 4.7) is the most frequently occurring littoral rissoid, and often the most abundant, with densities equivalent to tens of thousands per m^2 of habitat (Wigham & Graham 2017). It is not confined to rock-pool habitats, and can be found in silty situations amongst seaweeds and beneath stones across the entire mid and lower shore, and a wide range of wave exposure. *R. parva* breeds through much of the year. Mature females attach capsules containing up to 50 eggs onto any hard substratum; swimming veliger larvae hatch from these, settling after a planktonic life of a few weeks. Filamentous red seaweeds play an important role in the life cycle of this species, filtering settling larvae from the water column and providing surfaces on which to metamorphose, after which the juvenile snails disperse widely. As many as six cohorts, indicated by peaks in recruitment, may be apparent in any population of *R. parva*, but the individual life span does not exceed nine months (Wigham 1975).

veliger
a molluscan larval type that employs a ciliated bilobed flap of tissue, the velum, in swimming

cohort
in population ecology, refers to a generation arising from a single reproductive event

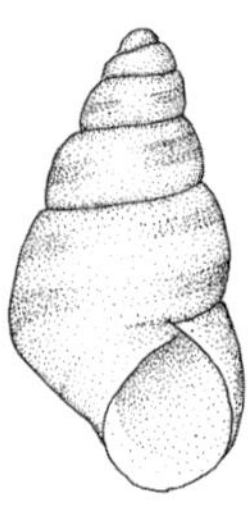

Fig. 4.8 The shell of *Barleeia unifasciata*. Scale bar: 2 mm.

Barleeia unifasciata (family Barleeidae) (Fig. 4.8) is a smaller but none the less conspicuous component of the algal-fringe fauna, although its geographical distribution is more restricted than that of *R. parva*, being limited to the south and west coasts of Britain and Ireland. Its tiny shell, up to 2 mm long, is a blunt, five-whorled spire, with a reddish-brown band on the lower half of each whorl.

The life cycle of *B. unifasciata* provides an interesting contrast to that of *R. parva* (Southgate 1982). It is principally associated with red seaweeds, occurring most frequently on filamentous species such as *Lomentaria articulata*, less frequently on *Osmundea pinnatifida*, which retain the detritus and diatoms on which the adult snails feed (Fig. 4.9).

The maximum individual life span of this species is around two years; gametogenesis appears to be continuous through most of the year but the size frequency distribution of a population at Bantry Bay, Ireland, showed two recruitment peaks, suggesting two distinct cohorts (Southgate 1982). Mature females attach their egg capsules to the algae on which they live. Each contains one egg, rarely two, which hatches as a crawling juvenile snail onto its preferred habitat; the hazard of a planktonic stage is thereby avoided, and recruitment to the population does not rely on physical entrapment. *Eatonina fulgida* (family Cingulopsidae) is even smaller than *B. unifasciata*; its similarly banded shell has

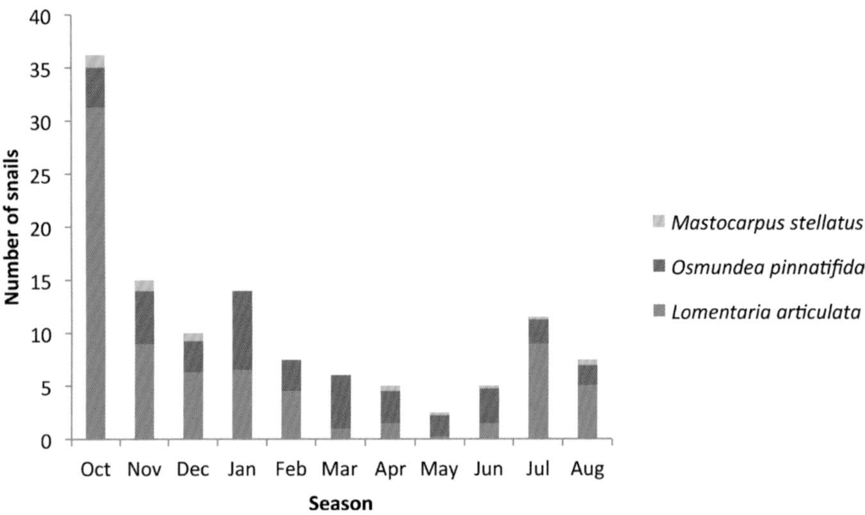

Fig. 4.9 Seasonal occurrence of *Barleeia unifasciata* on three seaweed species: numbers of snails per gram dry weight of weed. (Data from Southgate 1982)

just three whorls and a maximum length of 1 mm. It occurs over the same geographical range as the latter, in low-shore rock-pool habitats, and also attaches single encapsulated eggs amongst the *Corallina*/red algal fringe, which hatch as crawling juveniles, but the extent to which it is otherwise dependent on a specific host alga is unknown and requires investigation.

A pelagic phase in the life cycles of rocky-shore invertebrates facilitates dispersal and, importantly, has the potential to maintain gene flow between populations. In some species, fertilisation occurs in the water column and larvae develop within the plankton. In others, reproductive loss in these early stages in the life cycle is reduced by enclosing the fertilised eggs in capsules; these may be shed into the water column or glued to a firm substratum, in either case releasing pelagic larvae, which may spend from days to weeks in the plankton. In many littoral gastropods, larvae develop within a capsule attached to the substratum by the female snail, from which crawling juveniles hatch, a process termed 'direct development'. In herbivorous snails, egg capsules are often attached to the usual food algae, the juveniles being thus provided with a ready food source on hatching. Direct development involving encapsulation occurs in an especially advanced form among many whelks, such as the familiar Common Dog Whelk *Nucella lapillus*, in which the development of just a score or more embryos is fuelled by a store of non-developing 'food eggs' as a food source.

Brooded development, in which the entire developmental cycle is passed while the embryos are attached to or contained within an adult, is seen in numerous intertidal macrofaunal animals, including isopods, amphipods, some polychaetes and many shelled gastropods. The Rough Periwinkle *Littorina saxatilis* is the best-known example: fertilised eggs develop within the brood pouch of the female and are released as active, crawling, juvenile snails. Species lacking a pelagic dispersal stage may nonetheless display genetic differentiation between populations, which often show a gradient, or cline, of difference that increases with distance, a phenomenon referred to as 'isolation with distance'. Such gradients may be influenced by physical environmental factors, such as tides or currents, or unsuitable substrata, such as soft sediments, which might act as barriers to migration. Egg capsules and brooding adults might also achieve wide dispersal through rafting on drift algae; this might equally result in genetic connectivity between populations, or genetic differentiation as when only a subset of the parent genome establishes a new population.

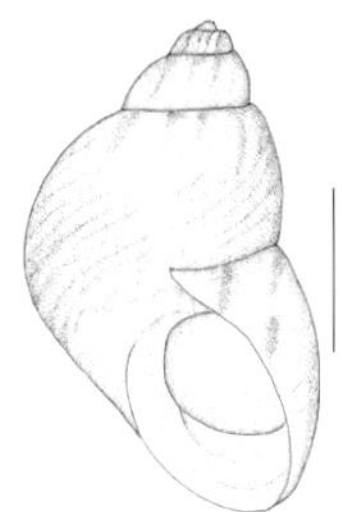

Fig. 4.10 The shell of *Tricolia pullus*. Scale bar: 2.5 mm.

Tricolia pullus (family Phasianellidae) is probably the most conspicuous and most readily recognised of the small epiphytic gastropods (Fig. 4.10); its boldly patterned shell may reach 10 mm in length, and the aperture is closed by a thick, white operculum. It is less common than *R. parva*, occurring in pools low on the shore and most often amongst small red seaweeds. Females lay eggs singly; these hatch within hours as veliger larvae with a very short free-swimming period (just a few days), which settle on the adults' host alga. It is not clear whether juvenile recruitment depends upon filtration, as with *R. parva*, or on substratum selection on the part of the settling larvae. It is perhaps possible that the unusually short development time, from hatching to settlement, might limit dispersal from the juveniles' optimum habitat. *Cerithiopsis tubercularis* (family Cerithiopsidae) (Fig. 4.11) and *Marshallora adversa* (family Triphoridae) are rather similar, with ridged, tuberculate shells, and a maximum length of around 6.5–7.0 mm. Both are associated with the sponges on which they feed, and occur on the lower shore and sublittorally to 100 m depth. On Britain's south and south-west coasts they can be found on the sponges *Halichondria panicea* and *Hymeniacidon perlevis*, intergrown with *Corallina* and filamentous red seaweeds. *Cerithiopsis tubercularis* lays egg capsules within the sponge tissue; each contains up to 200 eggs, which hatch as veligers that apparently have a planktonic existence of several weeks (Wigham & Graham 2017). The reproductive biology of *M. adversa* is less well known, although its veligers also appear to have a lengthy planktonic existence. With such a wide dispersal potential, it is not clear how the settling larvae of either species locate their preferred habitat.

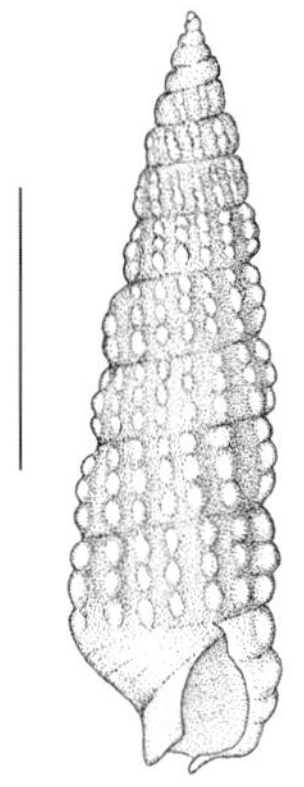

Fig. 4.11 The shell of *Cerithiopsis tubercularis*. Scale bar: 2.5 mm.

The smallest and least conspicuous of the shelled gastropods to be found in rock pools are *Skenea serpuloides* (family Skeneidae), *Skeneopsis planorbis* (family Skeneopsidae), *Omalogyra atomus* and *Ammonicera rota* (both family Omalogyridae), and three species of *Rissoella* (family Rissoellidae). All have been recorded living amongst filamentous algae in mid- to low-shore pools, and are also stated to occur sublittorally, although there appears to be no firm information regarding the ecology or biology of subtidal populations. *Rissoella diaphana* and the two species of Omalogyridae seem to have wide geographical distributions on British and Irish coasts, but the remaining species are known predominantly from south-west and western coasts only. It is probable that all are under recorded, overlooked simply because of their small size. *Skenea*

Fig. 4.12 The shell of *Ammonicera rota*. Scale bar: 0.5 mm.

Foraminifera
a class of amoeba-like protozoans, abundant in all marine habitats; individuals secrete perforated calcareous shells, termed 'tests'

serpuloides (Chapter 5, G.18) resembles a minute topshell, just 2 mm high, while species of *Rissoella* are 1–2 mm high and resemble tiny, smooth-shelled rissoids; in both cases they could be mistaken for juveniles of their analogues. *Skeneopsis planorbis*, *O. atomus* and *A. rota* (Fig. 4.12) have planospiral (i.e. ram's horn-shaped) shells that could not be easily confused with other littoral snails, but are so very small that perhaps they are just not seen, or even mistaken for foraminiferan tests.

All of these 'microgastropods' are known to produce encapsulated eggs, few at a time, that hatch as crawling juveniles, perhaps suggesting that, for each species, habitat preferences might be narrower than suspected; the two species of Omalogyridae are sequential hermaphrodites, perhaps also indicative of limited habitat range. Hermaphroditism is characteristic of the Heterobranchia (see below), which includes the Omalogyridae among many families of tiny, shelled gastropods, and almost all species of which display narrowly limited ecological distributions. The lack of a pelagic larval stage, and probable limited dispersal as a consequence, might suggest significant genetic differences between discrete populations in all of these species.

Sea slugs, formerly dignified as the subclass Opisthobranchia, comprise a heterogeneous assemblage of, apparently or obviously, unshelled marine gastropods, now assigned to a number of orders or suborders within the (appropriately denoted) subclass Heterobranchia. They are not uncommon between the tides on moderately exposed rocky shores. Many species will be individually large animals, sometimes brightly coloured; most are predatory and usually found in association with their invertebrate prey, and tend to occur in pools large enough to support suitably large populations of the sponges, bryozoans and other sessile organisms on which they feed. The Nudibranchia is the largest order and consists of two suborders. Species of the suborder Cladobranchia are mostly predators of cnidarians, ranging in size from the anemone-eating *Aeolidia papillosa*, common on most coasts, to the large (up to 200 mm) and bulky *Tritonia hombergii*, which feeds on the soft coral *Alcyonium digitatum* and might be found beneath shaded overhangs close to ELWS (Fig. 2.5). Very many of this suborder, such as species of *Doto* (Fig. 4.13), *Flabellina* and *Cuthona*, are tiny, rather decorative animals, often inconspicuous among their hydroid prey, and will only be found in habitats where these flourish.

Fig. 4.13 The nudibranch *Doto fragilis*. (Photo: J.S. Ryland)

The suborder Doridina is an equally diverse group among which sessile animals are the most frequent prey

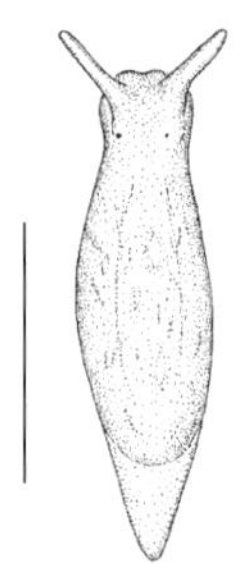

Fig. 4.14 The sacoglossan *Limapontia senestra*. Scale bar: 1 mm.

organisms: *Doris pseudoargus* (Chapter 5, K.15) is common in low-shore pools, where it feeds on the sponge *Halichondria panicea*; *Onchidoris muricata* (Chapter 5, K.18) is a predator of bryozoans; while *Gonidoris nodosa* feeds especially on colonial sea squirts, such as *Dendrodoa grossularia*. For most nudibranchs, all species of which display a pelagic larval phase, the presence of suitable prey seems to be the most important factor determining their occurrence in intertidal rock pools. However, for a few species of several smaller heterobranch groups, the algal fringes provide the optimum habitat. These are small, herbivorous animals, feeding upon their algal host plant, or grazing detritus, diatoms and microflora from its surfaces; they thus qualify as 'mesoherbivores'. Although actual diets are known with precision for relatively few species, most appear to feed by piercing cell walls of food alga species and sucking out cell sap.

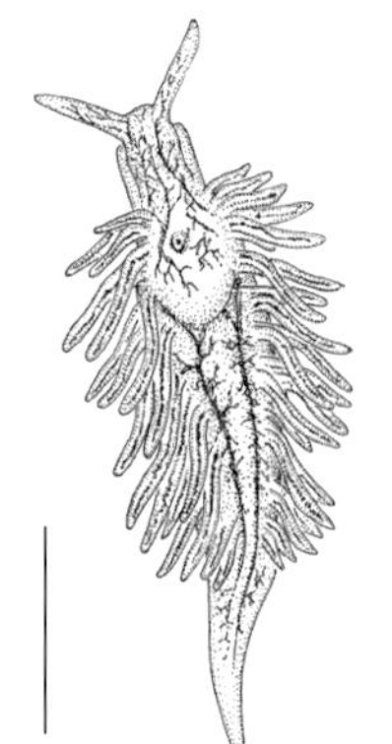

Fig. 4.15 The sacoglossan *Placida dendritica*. Scale bar: 2 mm.

On north-west European coasts the Sacoglossa encompasses the most familiar of these small sea slugs. *Limapontia capitata* and *Limapontia senestra* (Fig. 4.14) are rather nondescript animals, around 6 mm long, slender and tapered, distinguished from each other by an anterior pair of tentacles in the latter. They live on filamentous green algae, particularly *Cladophora rupestris*, and have interestingly contrasting life cycles. *Limapontia senestra* deposits its egg capsules onto its food plant, from which crawling juveniles hatch, while those of *Limapontia capitata* hatch as planktonic veligers; perhaps as a consequence, the latter can be found on a broader variety of filamentous green weed including *Chaetomorpha linum* and *Bryopsis* species, as well as *C. rupestris*. *Placida dendritica* (Fig. 4.15) and *Hermaea bifida* (Chapter 5, K.7) are two larger sacoglossans which might be found in low-shore pools on southern and western coasts. The former is associated with green algae, such as *Codium tomentosum* and *Bryopsis plumosa*, the latter with red algae, including *Halurus flosculosus*; in both species the life cycle includes a planktonic stage.

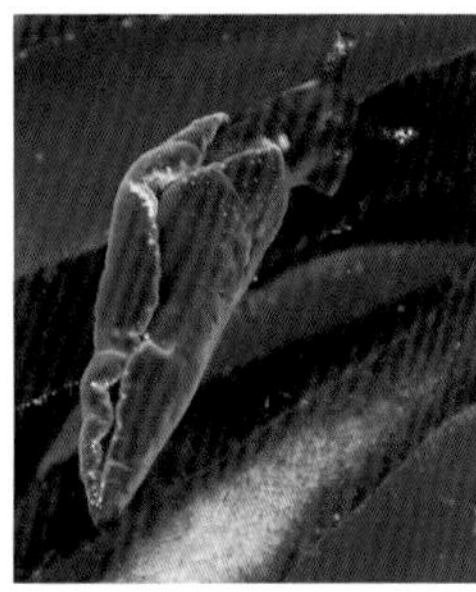

Fig. 4.16 The sacoglossan *Elysia viridis*.

Elysia viridis (Fig. 4.16) is the largest and most colourful sacoglossan found on north-west European shores; its green, brown or red body is flecked with red, blue and green spots. Its life cycle also includes a free-swimming veliger stage, and it also occurs on several species of large green algae, particularly fleshy coenocytic species.

coenocytic
lacking cell walls

Runcina coronata resembles a species of *Limapontia* in size, body shape and coloration but belongs to an entirely distinct order, the Runcinacea, and can be recognised by a protruding group of gills on its right postero-lateral side (Chapter 5, K.3). It is usually found in association with

Corallina officinalis and is thought to feed on the alga or its accumulated coatings of detritus, diatoms and microalgae; there is no free-swimming stage, but little is otherwise known of its biology or ecology.

4.4 Cushion stars and brittle stars

Two species of cushion star, *Asterina gibbosa* and *Asterina phylactica*, and the tiny brittle star *Amphipholis squamata* are part of the intertidal rocky-shore fauna of north-west Europe. The two cushion stars appear to be closely related (Darrock 2011) and have only been recognised as separate species comparatively recently (Emson & Crump 1979). They are readily distinguished by size and coloration: *A. phylactica* achieves a maximum diameter of 15 mm and has a dull greenish colour, with a central brownish, star-shaped pattern, while *A. gibbosa* is larger (up to 60 mm diameter), with coloration varying from light brown to orange, yellow or green, but with no trace of a dark pattern at any size. The two species show distinct differences in their ecological characteristics and reproductive biology (Crump & Emson 1983; Emson & Crump 1984). *Asterina gibbosa* occurs in rocky habitats from the lower shore, around ELWS (Fig. 2.5), and into the shallow sublittoral. Within the littoral zone it may range upshore to MHWN in large, deep pools, on both sheltered and wave-exposed coasts, typically on the lower sides of, or underneath, large rocks, in damp crevices and below shaded overhangs. *Asterina phylactica* has been reported in shallow coastal waters, to 18 m depth, but is principally distributed within the littoral zone, on wave-exposed shores, in rock pools around MHWN; its lower distribution is perhaps positively related to water flow and high-energy environments. Unlike *A. gibbosa*, the smaller *A. phylactica* does not shelter beneath rocks, occurring on the tops and sides of boulders, and may be very abundant in *Corallina* turf. *Asterina gibbosa* is a protandrous hermaphrodite (i.e. male first), switching from male to female at about two years of age, while *A. phylactica* is a simultaneous hermaphrodite and possibly self-fertilising. Both species form breeding aggregations, but while *A. gibbosa* groups subsequently disperse, leaving gelatinous egg masses glued to the substratum, the latter remain, covering and effectively brooding, or guarding, their eggs until hatching; both lack a pelagic larval stage, the eggs hatching as benthic juveniles.

Asterina gibbosa can be found on southern and western coasts of Britain, and all coasts of Ireland, but *A. phylactica* is presently known from just a few exposed shores, from south-west England, South Wales and the east coasts of Ireland to the Isle of Man. They are phylogenetically close, but molecular DNA studies indicate no evidence of hybridisation, even between populations of the two species sharing the same rock pool (Darrock 2011). Newly hatched juveniles are less than 1 mm in diameter and have been observed floating, upside down, at the water/air interface, a behaviour that probably results in small-scale dispersal. Longer-range dispersal almost certainly depends on algal rafting, although actively migrating adults perhaps achieve modest local range extension. Whatever the mode, dispersal promotes gene flow between local populations of both species; while genetic differentiation between populations increases with distance, there is no evidence that the brood-guarding behaviour of *A. phylactica* results in greater genetic diversity within or between populations than in *A. gibbosa* (Darrock 2012).

Amphipholis squamata is a very small brittle star, with disc diameter rarely exceeding 4 mm, and with slender arms up to 20 mm long. It is common to abundant on all rocky coasts of north-west Europe; it has been reported from shallow sublittoral habitats, and within the intertidal zone may range as far upshore as MHWN. It is essentially cryptic in habit, occurring beneath stones, in rock crevices and kelp holdfasts, and amongst fine-structured seaweeds, especially *Corallina officinalis* and red algal turfs. In suitable habitats *A. squamata* populations may be very dense: mixed turfs of *C. officinalis* and small red algae from a pool at MHWN on the South Devon coast yielded 74–300 individuals per 100 ml volume of alga (Emson & Whitfield 1989). *Amphipholis squamata* is hermaphroditic, and probably self-fertilising (Boissin *et al.* 2010), and is a brooding species, each individual brooding 1–6 embryos, depending on disc size; the South Devon study showed that reproduction occurred throughout the year, with a summer peak, and that brooding individuals were present in all months, though the proportion in the population was very low in the winter months. The species has been recorded from coastal habitats throughout the world, with the exception of polar seas, but with dispersal probably dependent on algal rafting, and potential gene flow between populations probably very low, such a widespread distribution suggests a complex of morphologically similar species. A rafting mat of *Corallina* and filamentous red seaweeds might carry several hundred *A. squamata*, which may then found a new population. Significant genetic differences have been found between individuals from populations separated by as little as 1 m, and even from a single sample, and two

clade
or 'monophyletic group', a group of genetically related organisms thought to comprise a common ancestor and its lineal descendants

distinct, non-hybridising clades have been recognised in populations from Roscoff, in Brittany, France (Le Gac *et al.* 2004, and references therein). Analysis of mitochondrial DNA from 64 specimens in a single sample of *Corallina* turf collected at Marseille, France, revealed five distinct genetic lineages within the two clades; nuclear DNA data indicated that four of these were reproductively isolated and could thus be regarded as biological species (Boissin *et al.* 2010). That individuals of five close genetic lineages are able to coexist in the same microhabitat remains to be explained, as competitive pressures might be expected to lead to ecological separation. Perhaps studies of the population dynamics and reproductive cycles of each may show how such a community is enabled to persist. Unfortunately, at present there are no morphological features that allow each 'species' to be distinguished.

4.5 Meiofaunal ostracods

The meiofauna is a significant component of algal fringe communities, consisting principally of nematode worms, mites and the smallest crustaceans. Among the latter group, harpacticoid copepods usually predominate, but small species of ostracod, commonly less than 1.0 mm long, may be almost as abundant, and equally diverse, and provide good examples of patterns in relative abundance and distribution probably common to meiofaunal algal communities. Monthly samples from a large (10 × 4.5 m), mid-shore rock pool on the Yorkshire coast, over a one-year period, revealed a fauna of 18 species of ostracod from four species of seaweed, namely *Ceramium virgatum*, *Chondrus crispus*, *Cladophora rupestris* and *Corallina officinalis* (Hull 1997). Species composition varied seasonally; eight of the ostracod species recorded were present in every monthly sample, in numbers sufficient to provide insights into their population cycles, while the rest were of only sporadic occurrence. The total density of the ostracod communities varied in relation to the seaweed species, and with season (Fig. 4.17).

Through the year, ostracod densities were consistently lowest on the broad-fronded, open-structured *Chondrus crispus*, and greatest among the filamentous tufts of *Ceramium virgatum*. *Cladophora rupestris* has a more richly branched structure than both *Chondrus crispus* and *Corallina officinalis*, and supported significantly higher densities of ostracods, but densities recorded on *Ceramium virgatum* were significantly greater than those on the other three algae. Seasonal variation in ostracod densities, and the composition of

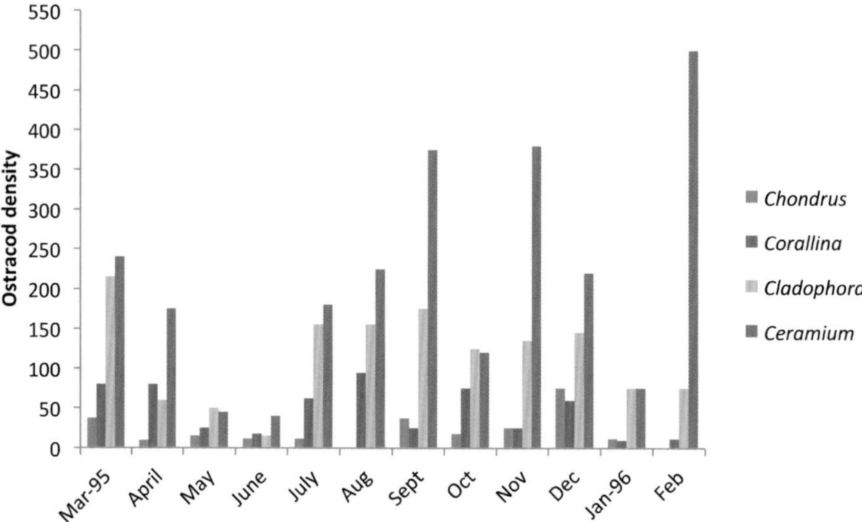

Fig. 4.17 Ostracod density in four rock-pool algae through 12 monthly samples, as numbers of individuals per 30 grams dry weight (*Corallina*) or per 15 grams dry weight (*Chondrus*, *Cladophora*, *Ceramium*). (Data from Hull 1997)

algal assemblages, also reflected the reproductive cycles of individual species (Fig. 4.18).

instar
in arthropods, a juvenile, developmental stage between successive moults

Cythere lutea reproduced in the early spring; the population peaked in March to April, when it consisted largely of the earliest larval instars, dropped sharply by late summer, and overwintered at low density as adults or last-stage instars. The *Hirschmannia viridis* population overwintered at a high density of late-stage instars, which moulted to adults and reproduced from March to May, and showed a peak in juvenile stages through July and August. *Xestoleberis aurantia* also overwintered as late-stage instars, but maturation and reproduction were more protracted, and peak numbers in September comprised both subadult and early-stage instars. Autumn and early-winter densities of *Boreostoma variabile* populations were significantly higher than in other months, and consisted mostly of late-stage instars and adults, with few juveniles recorded. This species appeared to reproduce throughout the sampling period, and population peaks might have represented overlapping generations and differential rates of maturation.

Total density, individual species abundance and community composition of algal-associated, rock-pool ostracod assemblages also display significant variation in relation to tidal level, and between different shores (Hull

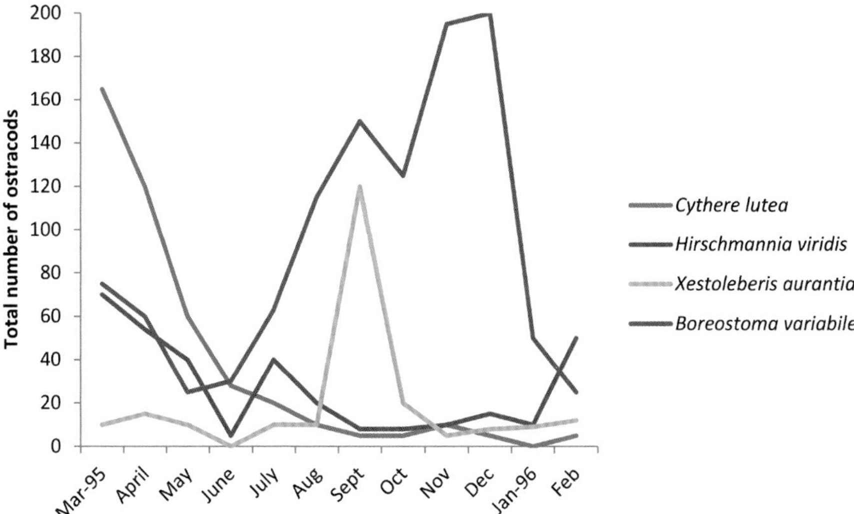

Fig. 4.18 Total monthly numbers of four ostracod species, combined for all four algal species. (Data from Hull 1997)

1999a). Of eight common species sampled in upper-, mid-, and lower-shore pools at four shores on the Yorkshire coast, six showed significantly different abundances between sites (Fig. 4.19).

For all species, density differed significantly in relation to tidal level (e.g. Fig. 4.19: compare A upper-, mid-, and lower-shore rock pools), while for seven there were significant differences between the four shores at equivalent tidal levels (e.g. Fig. 4.19: compare the upper-shore rock pools of A, B and C). The number of species and total abundance of individuals varied significantly between shores, although for each shore the number of species showed no significant difference between all samples.

Ostracod populations in a single species of rock-pool alga, *Corallina officinalis*, showed varying patterns of abundance between pools at a single tidal level on three of the Yorkshire sites (Hull 1999b). Five mid-shore pools sampled at each site yielded a total of 19 ostracod species; 14 occurred on all three shores, two on two of the shores while three were present on just one shore. The total number of individuals per 25 grams dry weight of alga did not differ significantly between shores, between pools or between samples within pools, but abundances of the nine most common species varied at different scales. For five species, mean abundance did not vary significantly between shores, but was significantly

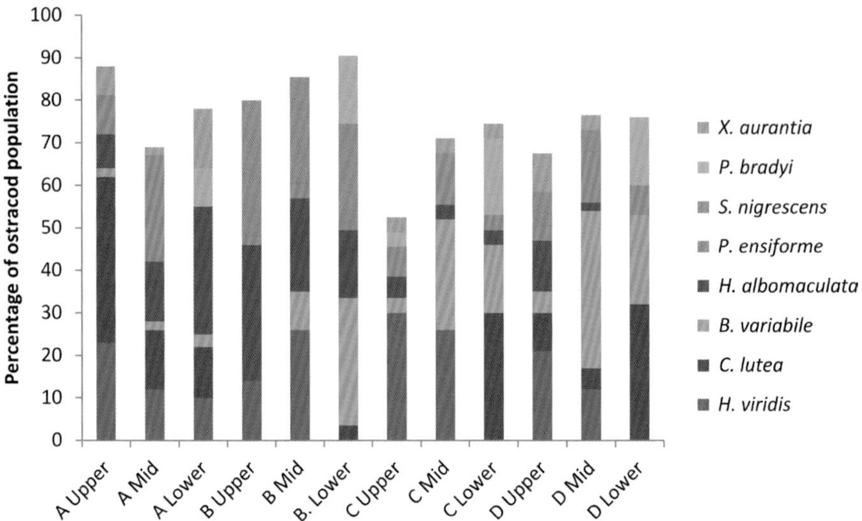

Fig. 4.19 Relative occurrence of eight ostracod species, in upper-, mid-, and lower-shore rock pools at four sites on the Yorkshire coast. A: Selwick Bay; B: Filey Brigg (sheltered); C: Filey Brigg (exposed); D: Ravenscar. (Data from Hull 1999a)

different between pools within each shore. For two species, abundance varied significantly between shores and pools, but not between samples from within each pool, while for another there was significant variation between shores, and between samples from within each pool, but not between pools, and for *Hirschmannia viridis*, the only significant variation was between samples taken from opposite sides of the same pool. Finally, the number of species recorded per quadrat differed significantly between all three localities, but not between pools, while species diversity differed significantly between both shores and pools, but not within pools.

Patterns of abundance and composition of algal ostracod communities are thus seen to be 'patchy' (Hull 1999b), varying seasonally, in relation to microhabitat (algal species), and at several spatial scales: between samples from a single pool, between pools and between shores. Low-shore assemblages tended to support lower numbers of species, and show least difference in composition between shores, while species composition and relative abundance showed significant differences between shores at higher tidal levels. Each rock pool is a unique habitat and such patchiness in ostracod assemblages might be attributable to several factors. Algal complexity is considered to be significant. Meiofaunal communities occur at highest densities amongst algae with complex frond structure, and at low density on least complex algae, perhaps reflecting small-scale microhabitat differences. Larger-scale patchiness might be related to physical factors, such as tidal level and wave exposure, or to differential recruitment and mortality, or local population extinctions.

5 Identification

The fauna and flora of rocky intertidal seashores are rich in species, any of which might be encountered in pools during low tide. The lower on the shore and the larger the pool, then the greater the array of littoral organisms to be found. However, as outlined in the preceding chapters, relatively few of these species will be limited to rock-pool environments. It is not possible to provide here identification keys to even a modest selection of the intertidal animals and plants encountered on the seashores of Britain and Ireland, and so recourse must be had to a good guide for the identification of the bulk of the species. An excellent source is *A Student's Guide to the Seashore* (Fish & Fish 2011), which offers a sound introduction to the ecology of intertidal habitats, including estuarine and coastal, hard and sedimentary environments, together with illustrated keys for the identification of plant and animal species, and notes on the biology and ecology of the most common and important species. A pocket guide might be useful for preliminary identification in the field. There are a number of these available, but it must be borne in mind that the Latin names of many genera and species, and their classification, may have changed in recent years as a consequence of molecular genetic research and a single taxon may bear different names in books just a year or two apart.

It is often desirable, when preparing reports or submitting species records, to check the current taxonomic status of each species identified online in the *World Register of Marine Species* (WoRMS). A more extensive guide to the marine fauna of the region is provided by the *Handbook of the Marine Fauna of North-West Europe* (Hayward & Ryland 2017), which describes and illustrates a comprehensive selection of intertidal and shallow subtidal organisms, from sponges to fish, together with dichotomous keys for their identification.

The keys provided in what follows are designed to allow the identification of the most frequently occurring and characteristic rock-pool animals on the shores of Britain and Ireland. The large brown algae distributed in distinct zones between the tidemarks, and the large foliose reds, are readily identifiable using a standard pocket guide. The most frequently occurring and ecologically significant filamentous red seaweeds, and the erect corallines, have been described and figured in Chapter 3; to identity additional species recorded during a survey (and there might be not a few!)

the most useful and practical source is *Seaweeds of Britain and Ireland* (Bunker *et al.* 2012), an especially user-friendly guide.

Key A Guide to major invertebrate animal groups

All animals collected are best examined in dishes of seawater, using a low-power stereomicroscope. It is often best to identify specimens while they are alive; indeed, many invertebrate species are characterised by colour patterns that fade rapidly after death. However, small crustaceans, apart from the larger isopods, usually have to be killed, by immersion in 70% methanol (Chapter 6; p. 156), and viewed by substage illumination, in order to be identified correctly.

modular
clonal animals forming colonies through continuous replication of functionally independent units – zooids, polyps etc. – from a single sexually produced larva

1. Sessile animals, solitary or colony-forming (modular). In white, calcareous, spiral tubes, firmly attached to algal substratum; ***or*** as spongy sheets or lobes; ***or*** encrusting; ***or*** erect, branching growths **2**
– Sedentary or freely moving animals, not permanently attached to a substratum. Solitary or aggregated but not colonial **3**

2. Tubeworms: occupying spiral tubes, to 2–4 mm diameter, cemented to coralline algae; when emerged, showing a fan of thick, stiff tentacles **Key J**
– Sponges (irregular bristly sheets, or flattened flask shapes, intergrown with small red algae), bryozoans and hydroids (encrusting or erect, branching growths, may evert circlets of delicate feeding tentacles when viewed underwater) **Key B**

3. Animals with jointed legs **4**
– Animals without legs **11**

4. With four pairs of long, spider-like legs, longer than body. Body slender, segmented; head with cylindrical proboscis (C.1) sea spiders (Pycnogonida) **Key C**
– With four pairs of short, clawed legs, shorter than the round or oval body (A.1) mites (Acari)

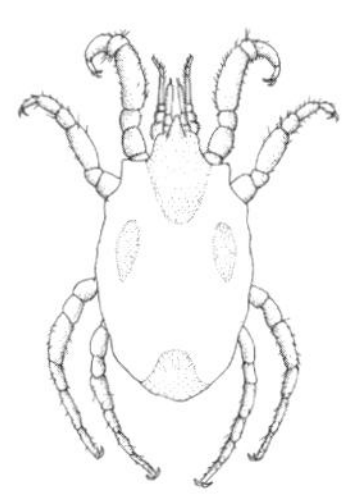

A.1 A marine mite, *Halacarellus basteri.*

Mites are not uncommon in the supralittoral zone of rocky shores, and some species occur in intertidal habitats. They are all very small (mostly around 1 mm long), often fast moving and sometimes brightly coloured. They are also very difficult to identify to species, requiring special preparation techniques and a high-power compound microscope. Bamber *et al.* (2017) tabulate morphological characteristics of the four suborders of Acari and provide diagrams illustrating salient features of 14 families likely to occur on the seashore. Green & Macquitty (1987) describe British species of the family Halacaridae.

– With **more** or **fewer** than eight legs **5**

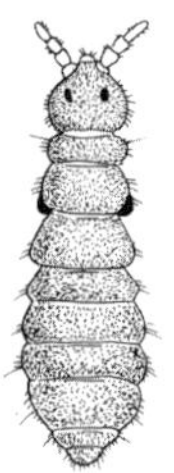

A.2 The collembolan, *Anurida maritima*.

5. Animal with six short, stumpy legs and two short antennae; body approx. 3 mm long, blue-black, distinctly segmented and hairy. Usually seen clustered in rafts on surface of upper-shore pools (A.2) *Anurida maritima*

This small arthropod belongs to the Collembola, a group closely related to insects and predominantly distributed in terrestrial habitats. It occurs at the very top of the intertidal zone and is best categorised as 'maritime' rather than 'marine'.

– With more than eight slender legs **6**

6. Legs all alike **7**

– Legs not all alike **9**

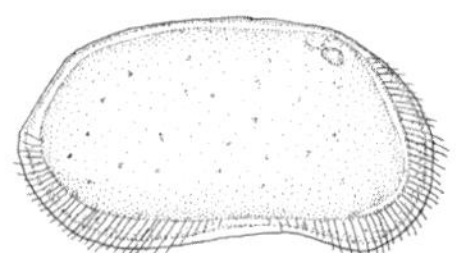

A.3 A marine ostracod, *Heterocythereis albomaculata*.

7. Small crustaceans, enclosed within oval or bean-shaped, bivalved shell, commonly <1 mm long; hinged dorsally, legs and antennae protrude ventrally (A.3) ostracods

Ostracods are small, mostly benthic, crustaceans, that are often common in silty habitats beneath stones, in rock crevices and algal holdfasts, and amongst *Corallina* and red algal turfs. While readily recognisable, they are difficult to identify to species. Samples may be sorted according to individual sizes and shell morphology, but specialist literature (e.g. Athersuch *et al.* 1989) is required to enable specific identification.

– Small or large, but not enclosed within a bivalved shell **8**

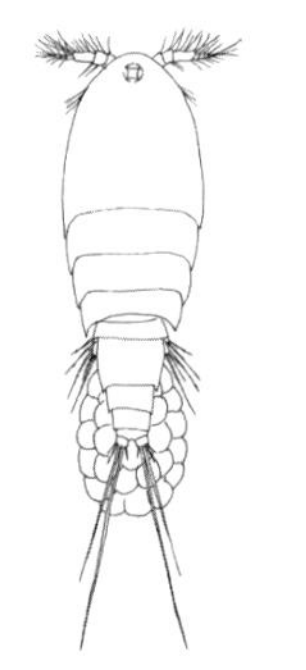

A.4 A copepod often found in rock pools, *Diarthrodes nobilis*.

8. Small crustaceans, commonly <5 mm long, with slender, fusiform body; very small legs at front end of body, hind end tapered to a pair of long or short processes, the caudal rami, usually bearing spines or setae (A.4) copepods

The subclass Copepoda is represented in the north-west European marine area by around 1,500 species, occurring in both benthic and planktonic habitats, and as commensals and parasites of most marine organisms. Species of the order Harpacticoida are common (often abundant) in silty littoral habitats, including rock pools. All are small; identification is difficult, requiring high-power microscopy and specialist keys. *Tigriopus brevicornis* (Fig. 4.4) is recognised by its bright orange coloration, and its unique supralittoral habitat. Huys *et al.* (1996) provide descriptions of further species, together with keys for identification.

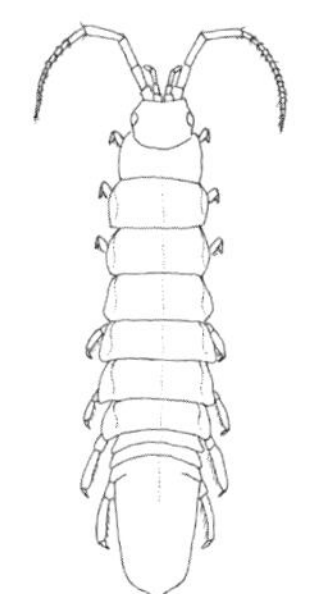

A.5 An isopod, *Idotea chelipes*; male.

– Larger crustaceans, >5 mm. Body dorsoventrally flattened, with distinct head, bearing two pairs of antennae, segmented middle region bearing paired, identical legs, and distinct tail portion (A.5) isopods **Key D**

9. Head and most of thorax enclosed by a flexible carapace; with a distinct rostrum projecting between a pair of large eyes; thorax with four pairs of biramous (two-branched) appendages, first two pairs usually modified for feeding, the rest for locomotion; abdomen with six pairs of appendages, each having a single peduncle with two short branches. Distinct tail fan with conspicuous pair of statocysts (A.6) Mysidae

   Mysids (often termed opossum shrimps) occur commonly in inshore waters, sometimes in large shoals, and several species are occasionally found amongst weed in large, low-shore pools. Common coastal species are described and illustrated by Ashton *et al.* (2017).

– Not as described **10**

peduncle
basal segment of appendage

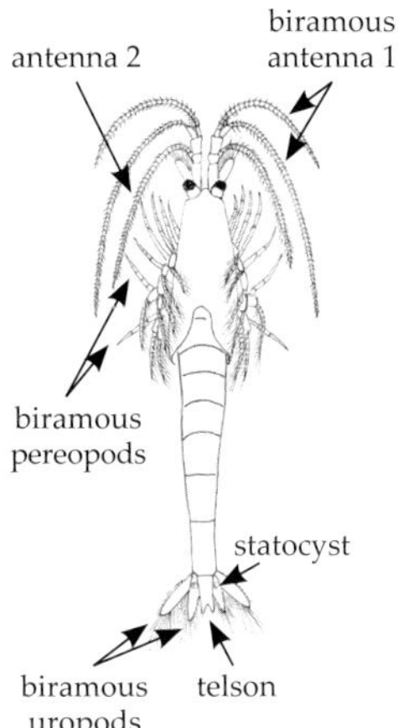

A.6 A mysid, *Gastrosaccus sanctus*.

10. Small crustaceans, mostly <10 mm long, lacking a carapace, laterally flattened. Typically with seven pairs of thoracic legs (pereopods), three pairs of abdominal legs (pleopods) and three pairs of posterior uropods (A.7) amphipods **Key E**

    The Caprellidae is an anomalous family characterised by a slim, cylindrical body comprising a bulbous head and seven body segments, and a reduced number of legs (A.8). Aptly described as 'skeleton shrimps', they are often found amongst hydroids, erect bryozoans and fine-structured seaweeds. Around 20 species are known from north-west European coasts.

– Decapods: shrimps, prawns and crabs. Head and thorax fused as the cephalothorax, enclosed by a laterally or dorso-ventrally flattened carapace; five pairs of thoracic legs (posterior pair may be tucked beneath carapace) and five pairs of abdominal legs (flexed beneath thorax in crabs) **Key F**

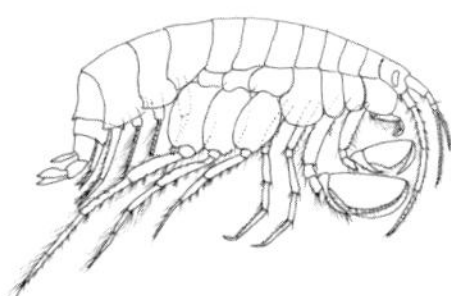

A.7 An amphipod, *Liljeborgia pallida*.

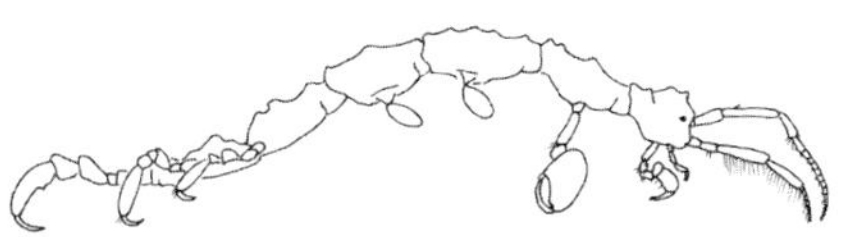

A.8 A caprellid amphipod, *Caprella acanthifera*.

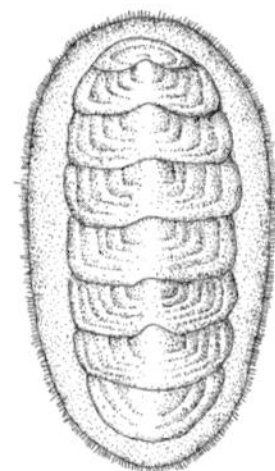

A.9 A chiton, *Boreochiton rubra*.

11. Animals with an external shell, entire (i.e. single) ***or*** constructed of a row of curved plates (A.9) ***or*** consisting of two dorsally hinged valves **12**

– Animals soft bodied, lacking an external shell **13**

– Sea urchins, starfish and brittle stars: stiff bodied, with an outer skeleton of small calcareous plates; urchins with radiating rows of spines, others with short rows or whorls of spines on arms **Key M**

12. Shell entire, conical (limpets) or spiralled (snails) shelled gastropods **Key G**

– Shell consisting of a series of curved, overlapping plates chitons **Key H**

– Shell of two valves, with dorsal hinge; opening ventrally to show animal's foot bivalves **Key I**

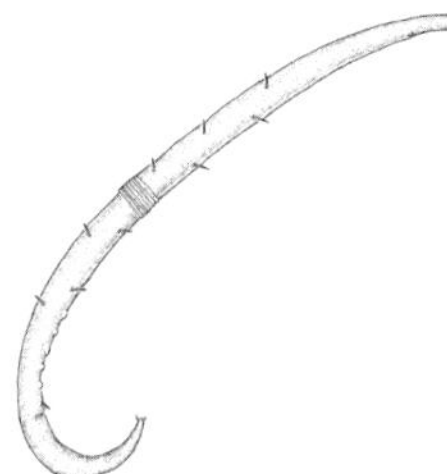

A.10 A generalised marine nematode. (After Platt & Warwick 1983.)

13. Worms; with bodies distinctly segmented ***or*** unsegmented and rather stiff ***or*** flat and ribbon like **14**

– Not worms **15**

14. Nematodes (round worms), with smooth, stiff, cylindrical bodies (A.10) and nemerteans (ribbon worms), with thin, flat, fragile bodies (A.11) Nematoda and Nemertea

Identifying species of these two phyla is difficult. Nematodes are ubiquitous as part of the marine meiofauna; all are small and nondescript. Platt & Warwick (1983, 1988) provide keys and descriptions for the orders Enoplida and Chromadorida, while Warwick *et al.* (1998) treat the order Monhysterida. Some nemerteans are large and distinctively patterned; Gibson & Knight-Jones (2017) provide keys and descriptions for common coastal species.

– Polychaetes (bristle worms): bodies distinctly segmented, each segment with paired appendages, bearing bundles of bristles, long cirri or paddles; head with eyes, antennae and cirri, ***or*** with feathery feeding tentacles; free living ***or*** occupying a permanent tube **Key J**

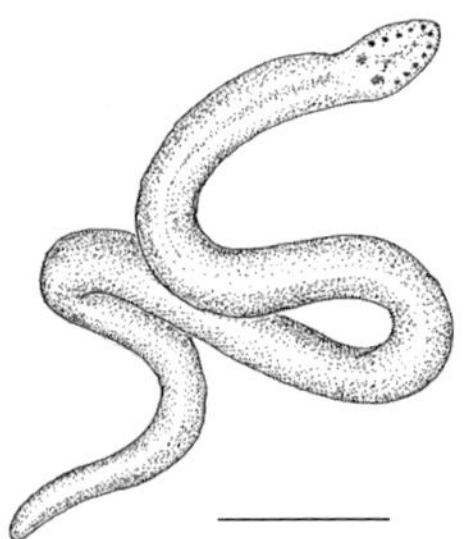

A.11 A nemertean, *Lineus ruber*. Scale bar: 5 mm.

15. Heterobranchia (sea slugs): bilaterally symmetrical gastropods, with a head bearing various tentacle-like processes; body with lateral flaps, or dorsal tubercles or longer processes **Key K**

– Sea anemones: radially symmetrical animals with whorls of tentacles around a central, upper mouth **Key L**

Key B Sessile, modular animals

Sponges, hydroids, bryozoans and ascidians (sea squirts) comprise the majority of the communities of sessile animals on intertidal rocky shores that encrust stable rock, large boulders and shell debris, as well as living substrata such as algae, gastropod shells and decapods. The diversity of sessile communities is related to tidal level, degree of wave exposure and shade, and is typically richest on shaded, overhanging surfaces on the lower shore. Many species of hydroid and bryozoan are associated with particular algal species, or are part of the kelp holdfast community, while perhaps a majority of sponges occurring between the tides are essentially epilithic in habit. The following key is designed for the identification of sessile animals associated with the coralline/red algal fringe, together with a few species of bryozoan common on brown algal fronds in mid-shore pools. A more comprehensive key to coastal sponge species is provided by Goodwin *et al.* (2017). All north-west European species of hydroid are described and figured by Cornelius (1995) and Schuchert (2012), while Hayward (1985) and Hayward & Ryland (1985, 1998, 1999) allow identification of all bryozoan species likely to occur in the region. Porter (2012) provides descriptions of more than 120 species of bryozoans and hydroids, with notes on their biology and ecology, each illustrated by colour images of live specimens; a similar coverage of sponges and sea squirts, also illustrated by colour photography, can be found in Bowen *et al.* (2018).

epilithic
encrusting rock

osculum (pl. oscula)
in sponges, opening for the excurrent water flow

1. Sponges: forming sheets, clumps or lobes, or erect flask shapes; surface smooth or bristly, with one or more conspicuous oscula, but lacking distinct, repeated morphological features **2**
– Hydroids and bryozoans: encrusting, creeping or erect; as sheets, stolons or branched, but clearly composed of distinct modules (polyps or zooids), each with a circlet of eversible tentacles **4**

2. Sponge in the form of a short, stiff cylinder or barrel; ***or*** a thin flat, sheet; ***or*** tubular and branched, forming a dense clump; not forming a continuous encrusting sheet. Rarely exceeding 1 cm high; oscula always single, terminal and conspicuous (B.2, B.3). Off-white or yellowish **3**
– Smooth-surfaced; oscula regularly spaced, with distinct, raised rims. Commonly green, sometimes yellow or orange. Widespread and common, on a range of substrata, on middle and lower reaches of rocky shores. Frequently densely intergrown with red algal turfs,

B.1 The encrusting sponge *Halichondria panicea*. (Photo: J.S. Ryland)

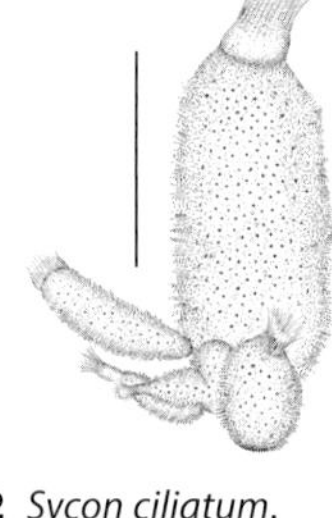

B.2 *Sycon ciliatum*. Scale bar: 5 mm.

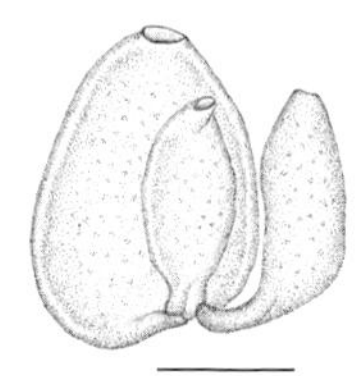

B.3 The Purse Sponge *Grantia compressa*. Scale bar: 5 mm.

and often developing extensive encrustations on kelp holdfasts (B.1) *Halichondria panicea*

3. Sponge vase shaped or cylindrical, the osculum surrounded by a fringe of stiff bristles. On *Fucus serratus*, and other low-shore algae, in sheltered habitats, and also among small red algae (B.2) *Sycon ciliatum*

– Sponge flat, oval or irregular, with a smooth-rimmed osculum. Widespread and common, on many low-shore algae; often abundant among small red seaweeds (B.3) Purse Sponge *Grantia compressa*

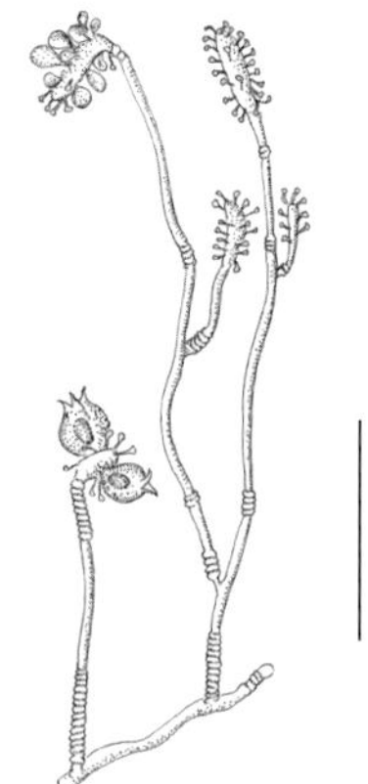

B.4 Erect stems of *Sarsia eximia*. Scale bar: 10 mm.

4. Hydroids (B.4–B.9): colonies formed of stalked polyps arising from a creeping stolon; ***or*** as erect branching forms, often in strictly ordered series. Colony with continuous horny outer wall (the perisarc); polyps bear slender, non-ciliated tentacles, and are linked by a common, colony-wide body cavity **5**

– Bryozoans (B.10–B.26): colonies formed of encrusting sheets, mounds or loose clumps, ***or*** erect branching forms; a minority as stalked zooids from an encrusting stolon. Outer body wall of zooids calcified or gelatinous. Orifice of each zooid closed by an opercular flap or muscular sphincter. Tentacles bear cilia – seen as a flickering iridescence on everted polyps **10**

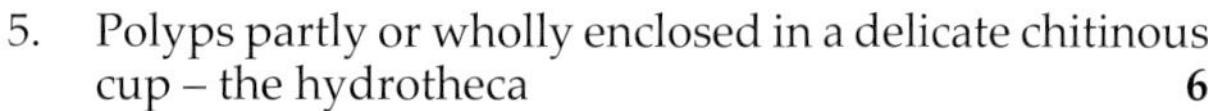

5. Polyps partly or wholly enclosed in a delicate chitinous cup – the hydrotheca **6**

– Polyps not enclosed by a cup. Colony with a slender, erect stem, branching irregularly, the polyps spaced along the stem and branches **7**

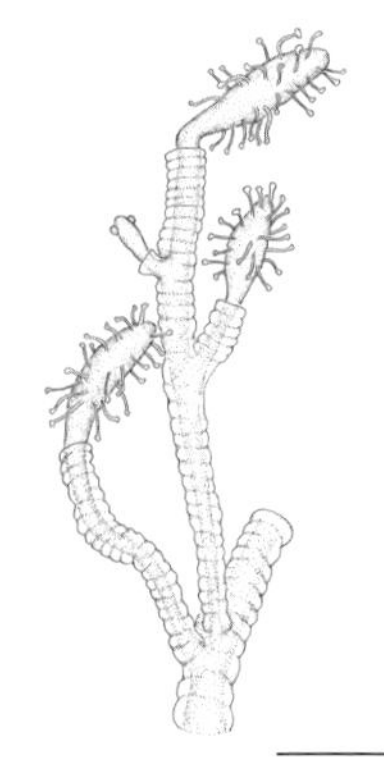

B.5 The hydroid *Coryne pusilla*, showing the ringed perisarc. Scale bar: 1 mm.

6. Perisarc with irregular groups of rings, separated by smooth, cylindrical portions. Up to 7 cm high; intertidal, in rock pools, frequently on algae (B.4) *Sarsia eximia*

– Perisarc irregularly ringed along its whole length, typically dark brown. Intertidal, in rock pools and often on algae (B.5) *Coryne* species

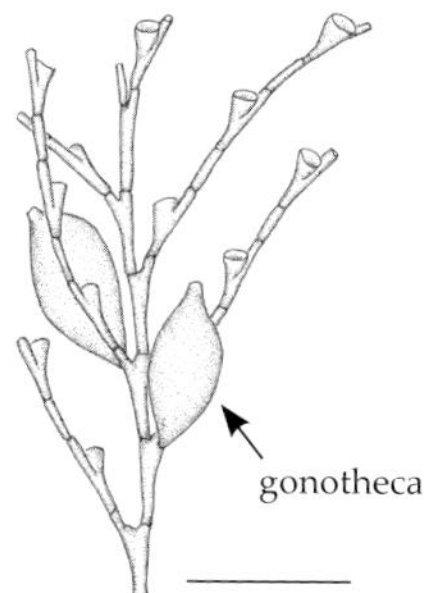

B.6 A branch of *Plumularia setacea*, with characteristic gonothecae. Scale bar: 1 mm.

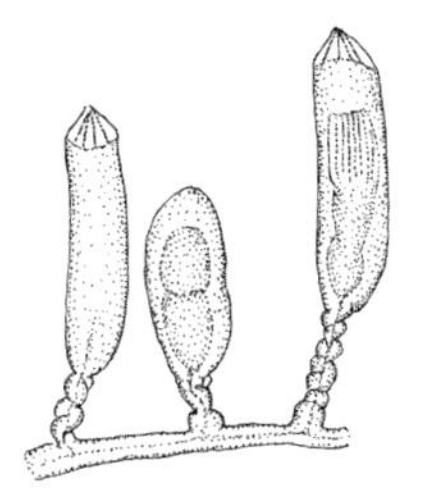

B.7 *Calycella syringa*: hydrothecae with caps, and a typical gonotheca. Scale bar: 0.5 mm.

gonotheca (pl. gonothecae) specialised theca (cup) enclosing an elongate process, the blastostyle, from which medusae are budded

7. Colony branching regularly to form a delicate feather shape. Hydrotheca deeper than wide, with smooth rim. Ovoid gonothecae situated in axils between main stem and side branches. Small, typically less than 5 cm. Lower shore, often in rock pools; on rock or large algae (B.6) *Plumularia setacea*

– Colony branching irregularly, not forming a feather shape; each hydrotheca attached to the stem of the colony by a short stalk **8**

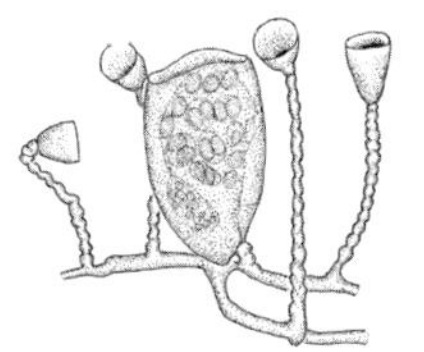

B.8 *Orthopyxis integra*: group of hydrothecae, and a typical gonotheca. Scale bar: 1 mm.

8. Hydrothecae elongate, cylindrical; each with a conical cap at the end (lost on death of polyp); short-stalked gonothecae interspersed between them. On hydroids and algae, lower shore and subtidal (B.7) *Calycella syringa*

– Hydrotheca bell shaped, without cap **9**

9. Hydrotheca with a smooth rim, its stalk with irregular rings; gonotheca large, urn shaped. On red algae, lower shore and subtidal (B.8) *Orthopyxis integra*

– Hydrotheca with toothed rim, its stem with groups of regular rings. Common intertidally on a range of algae (B.9) *Clytia hemisphaerica*

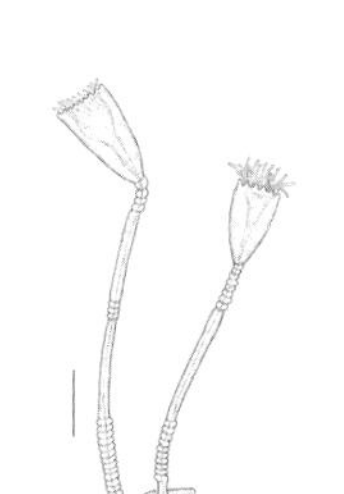

B.9 *Clytia hemisphaerica*: two typical hydrothecae. Scale bar: 0.5 mm.

10. Colonies encrusting; forming flat sheets, small patches or mounds, or sometimes larger, fleshy and knobbed growths from an extensive encrustation **11**

– Colonies not encrusting, erect or hanging; forming thick, bushy or spiralled growths up to 20 mm high, ***or*** diffuse and spindly, ***or*** in dense, tangled tufts, ***or*** with small (<1 mm), upright cylindrical zooids attached to a creeping stolon **17**

B.10 The epiphytic bryozoan *Flustrellidra hispida*. Note spines bordering the orifice. Scale bar: 0.5 mm.

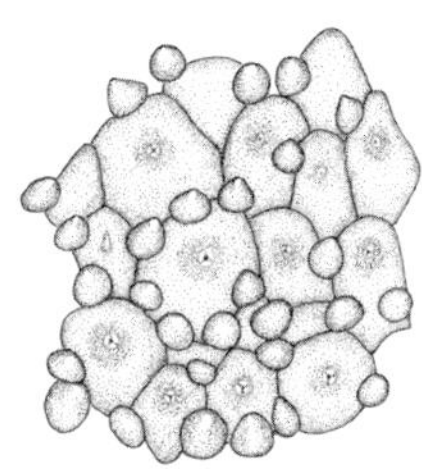

B.11 *Alcyonidium hirsutum*: group of zooids with kenozooidal papillae. Scale bar: 0.5 mm.

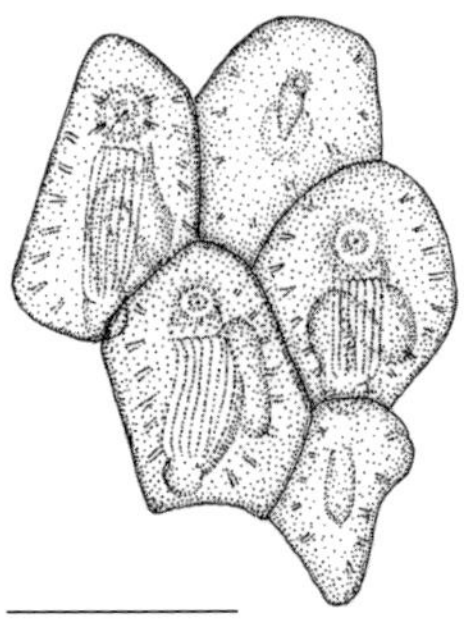

B.12 Flat, smooth-surfaced zooids of *Alcyonidium gelatinosum*. Scale bar: 0.5 mm.

11. Colonies uncalcified, smooth and gelatinous, or rather fleshy with numerous papillae on the surface, or with thick, brown spines **12**

– Colonies partly or wholly calcified; sometimes with the frontal surface of the zooid membranous, to a greater or lesser extent, but with at least the vertical walls of the zooids calcified **14**

12. Each zooid bearing few or many sharply pointed chitinous spines. Colony purplish brown. Predominantly on *Fucus serratus*; also *Mastocarpus* and *Chondrus* (B.10) *Flustrellidra hispida*

– Zooids without spines **13**

13. Colony surface with numerous rounded or conical papillae, visible with a hand lens; velvety to touch. Forms flat encrustations, or sometimes short, finger-like erect lobes. Predominantly on *Fucus serratus*, also *Chondrus* and *Gigartina*, more rarely on other small red algae (B.11) *Alcyonidium hirsutum*

– Colony surface flat, smooth to touch. Forms a thin, gelatinous incrustation on *Fucus serratus*, very rarely on other algae (B.12) *Alcyonidium gelatinosum*

14. Frontal surfaces of zooids partly or wholly membranous; sometimes with a border of spines, but always visible **15**

– Frontal surfaces of zooids calcified; smooth, rough or porous, but always with no visible membranous area **16**

15. Colony forming extensive, white, lace-like incrustations on the fronds of kelp. Zooids rectangular, minimally calcified, with short, conical knobs at each corner; frontal

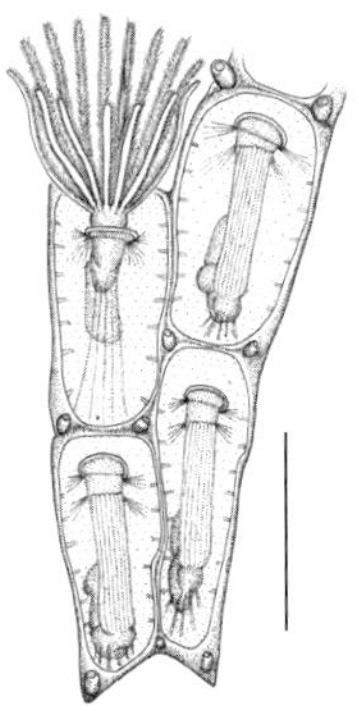

B.13 Zooids of *Membranipora membranacea*. Scale bar: 0.5 mm.

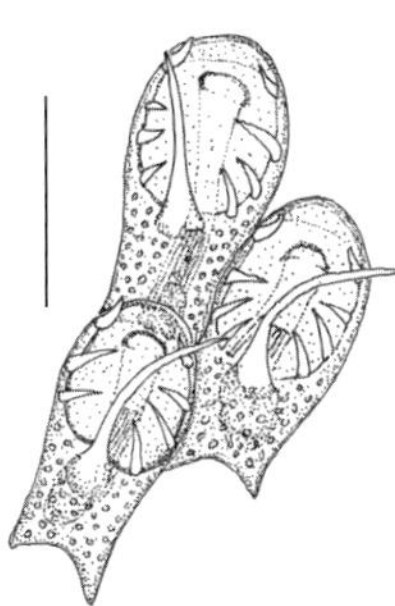

B.14 *Electra pilosa*: **a** colonies encrusting *Fucus serratus*; **b** group of three zooids. Scale bar: 0.5 mm.

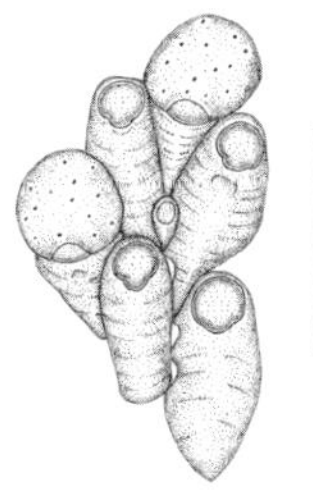

B.15 *Celleporella hyalina*: globular ovicells are borne by small female zooids. Scale bar: 0.5 mm.

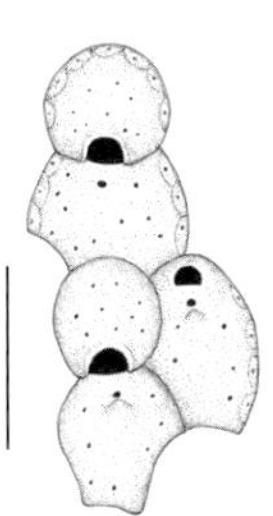

B.16 *Haplopoma impressum*: ovicells develop on normal-sized zooids. Scale bar: 0.5 mm.

surfaces entirely membranous with the withdrawn tentacles quite visible. On the fronds of *Laminaria*, rarely on *Fucus serratus*. Common on all rocky coasts (B.13) *Membranipora membranacea*

– Colony smaller; circular, star shaped or irregular, commonly less than 1 cm^2. Frontal surface of zooid with a membranous portion, surrounded by 4–12 spines, and a calcified portion with numerous pores. On a wide range of lower shore algae, common on all British coasts (B.14) *Electra pilosa*

16. Zooids of several different sizes, ovicells (larval brood chambers) borne only by small, specialised female zooids. Orifice of zooid with concave lower edge. Frontal walls without pores. On *Laminaria* holdfast, stipe and frond, also on numerous small red algae. Widespread and common (B.15) *Celleporella hyalina*

– Zooids of one size only, ovicells borne by ordinary zooids. Orifice of zooid with straight lower edge. Frontal wall with scattered pores. On small red algae, south and west coasts only (B.16) *Haplopoma impressum*

17. Colony calcified, white, delicate; comprising chains of horn-shaped zooids in single rows, ***or*** a slender, cylindrical stolon bearing individual, straight or curved,

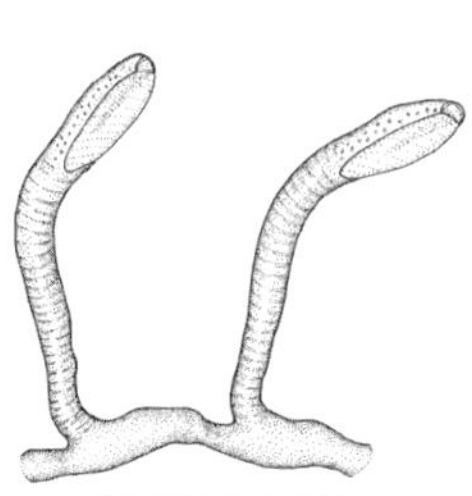

B.17 Two zooids of *Aetea anguina*; each comprises an erect, tubular portion, with downcurved, spoon-shaped anterior, and an encrusting base. Scale bar: 0.5 mm.

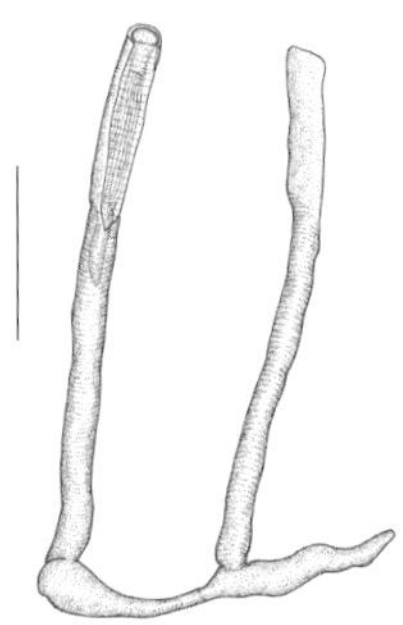

B.18 Two zooids of *Aetea sica*, rising from encrusting basal portions. Scale bar: 0.5 mm.

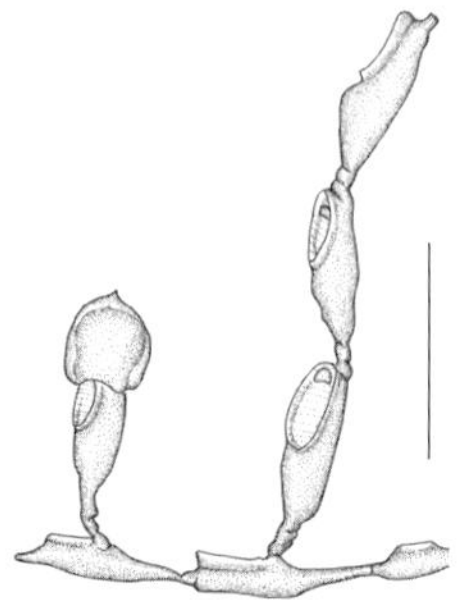

B.19 *Scruparia ambigua*: encrusting zooids, giving rise to an erect chain of zooids and a single brooding zooid. Scale bar: 0.5 mm.

upright cylinders, each with a membranous region at the end **18**

– Not as described **19**

18. Zooids in branching chains creeping over the substratum, or occasionally partly detached and forming a dense tuft. Each zooid comprising an encrusting portion and an erect, cylindrical upper portion (B.17–B.18). On lower shore algae, particularly *Laminaria* and small red algae, also on other bryozoans, and hydroids:

– End of zooid spoon shaped and downcurved (B.17) *Aetea anguina*

– End of zooid straight, cylindrical (B.18) *Aetea sica*

– Zooids in branching chains, at first encrusting but developing erect, tangled growths; horn shaped, with the membranous frontal surface at the broad end. On kelp holdfasts and small red algae, often on other bryozoans and hydroids (B.19) *Scruparia* species

B.20 *Crisidia cornuta*: **a** branch showing joints (nodes) between each zooid; **b** the gonozoid in lateral and frontal views. Scale bar: 0.5 mm.

19. Colony calcified, branching, with brown, black or colourless joints between branches; zooids tubular, with rounded orifices at their ends, budded from preceding zooids, without connecting stolons **20**

– Colony uncalcified, translucent, although often with adhering detritus. Zooids cylindrical, with orifices at their tips, puckered when closed; budded singly or in groups from a cylindrical stolon (B.22–B.26) **21**

20. Each internode (i.e. the portion of the colony between two joints) consisting of a single zooid bearing a long, pointed spine. On kelp stipes and small red algae (B.20) *Crisidia cornuta*

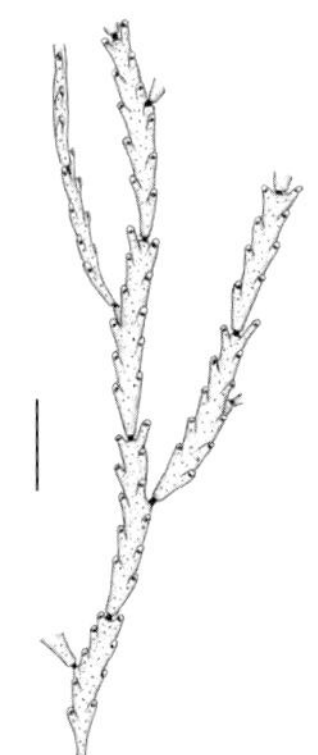

B.21 *Crisia denticulata*: part of a colony, long internodes separated by dark, chitinous nodes. Scale bar: 1 mm.

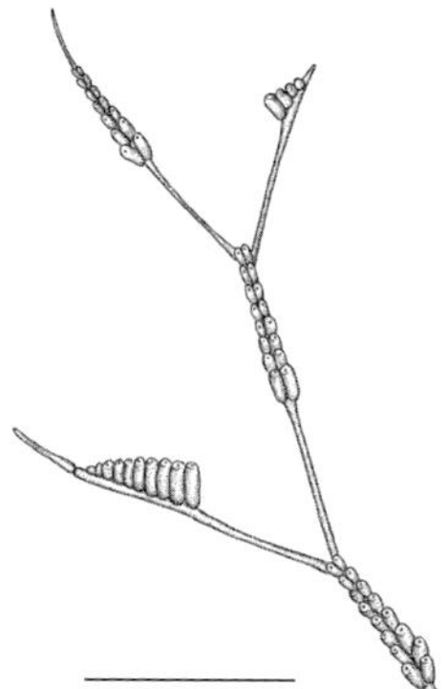

B.22 *Amathia lendigera*: part of colony with separate comb-like rows of zooids. Scale bar: 2 mm.

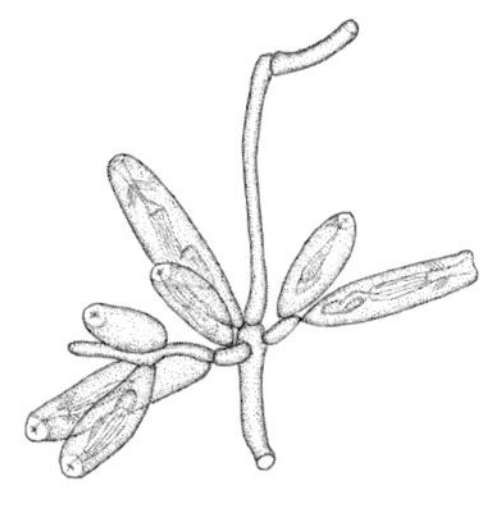

B.23 *Walkeria uva*: zooids and kenozooids on a portion of the stolon. Scale bar: 0.5 mm.

– Each internode consisting of two to many zooids, with or without spines. On kelp stipes and small red algae. Frequently forming a dense, luxuriant turf (B.21) *Crisia* species

21. Colony stiff, brown, tufted or diffuse. Zooids arranged in short, doubled panpipe-like series. On kelp holdfasts and stipes, and amongst bryozoan turfs (B.22) *Amathia lendigera*

– Colony diffuse or tufted, ***or*** forming a dense sward. Zooids single, clumped or spiralled **22**

22. Zooids in clumps or fans, budded from groups of tiny, rectangular kenozooids, linked by a slender stolon. On *Corallina officinalis*, small red algae, kelp holdfasts and amongst bryozoan turfs (B.23) *Walkeria uva*

– Zooids budded directly from the colony stolon. Colony creeping, in hanging tufts, or developing a thick mat, but without regular leaf-like arrangement of zooids. On *Fucus serratus*, particularly in sheltered estuarine areas, and on other low-shore algae **23**

kenozooid
a specialised zooid lacking orifice, tentacles and gut, simply a coelomic space bounded by outer body wall

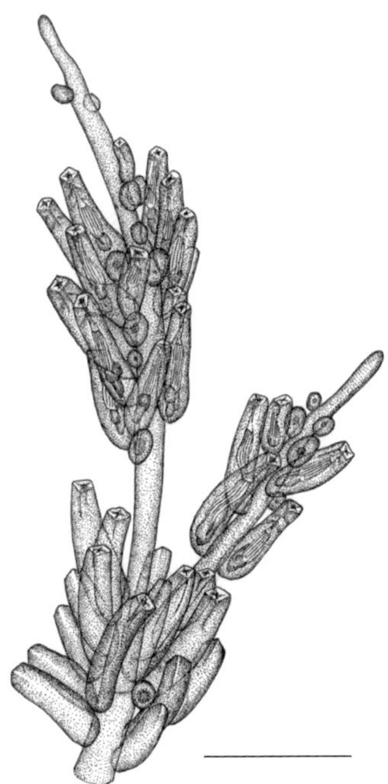

B.24 *Amathia imbricata*: groups of overlapping zooids spaced along the stolon. Scale bar: 0.5 mm.

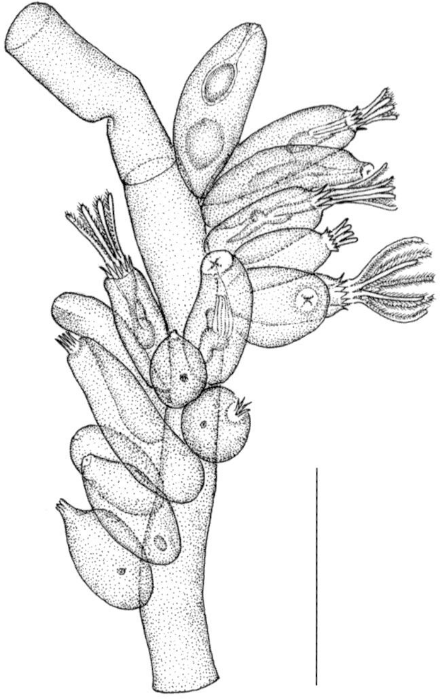

B.25 *Amathia citrina*: zooids spiralled around the stolon. Scale bar: 0.5 mm.

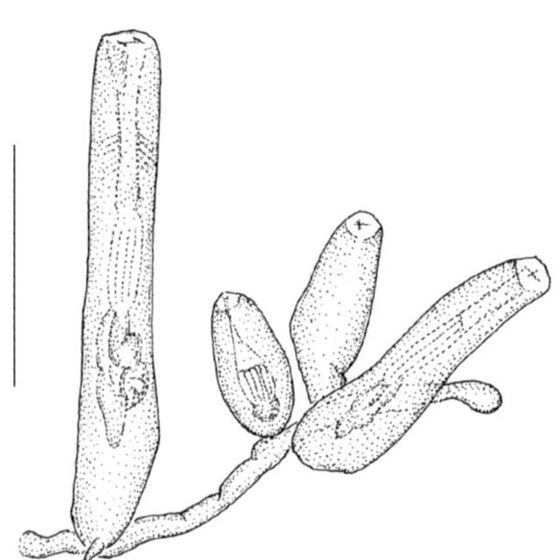

B.26 *Amathia gracilis*: zooids irregularly spaced along the stolon. Scale bar: 0.5 mm.

23. Dense swards of overlapping zooids on *F. serratus* (B.24) *Amathia imbricata*
– Hanging tufts. Zooids in loose spirals around the stolon *Amathia pustulosa*
– Hanging tufts. Zooids in spirals, bright lemon colour (B.25) *Amathia citrina*
– Diffuse, creeping colonies. Zooids in small dense clusters, or single (B.26) *Amathia gracilis*

Key C Sea spiders (Pycnogonida)

All species of pycnogonid recorded from the north-east Atlantic region, from the shore to the continental slope, are described and illustrated by Bamber (2010), together with notes on their biology and ecology, and keys for their identification. The few species keyed out here are those which are presently known to occur commonly in red algal turfs or as micropredators of sessile epiphytes.

1. Chelifores and palps present (C.1) **2**
– Chelifores or palps, or both, lacking **6**

2. Palps five-jointed. Chelifores longer than proboscis **3**
– Palps with eight or nine joints. Chelifores shorter than proboscis **4**

3. Last two segments of palps of equivalent length (C.2). Body smooth, up to 8 mm long, with relatively elongate proboscis. Legs up to 30 mm long, smooth, with a few bristles on the last segment. A carnivore, known to feed on hydroids (*Dynamena pumila*) and bryozoans (*Amathia* species) *Nymphon gracile*
– Last segment of palp about twice length of preceding segment (C.3). Body up to 5 mm long, proboscis relatively short and stout. Legs up to 20 mm long, with

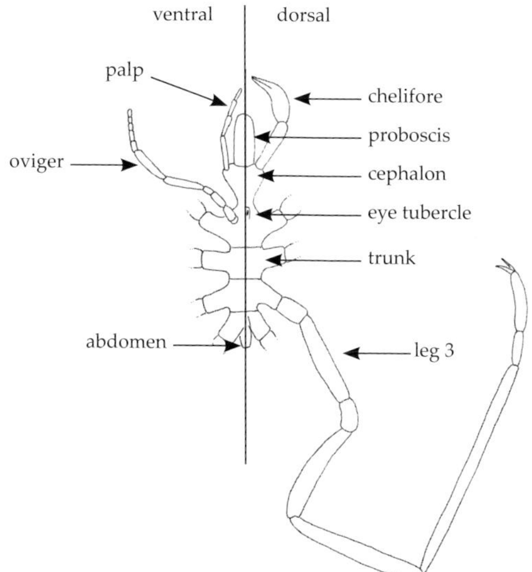

C.1 Structure of typical pycnogonid.

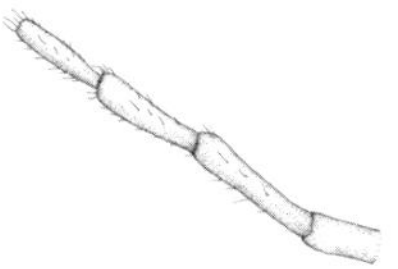

C.2 Distal portion of palp of *Nymphon gracile*. Scale bar: 0.5 mm.

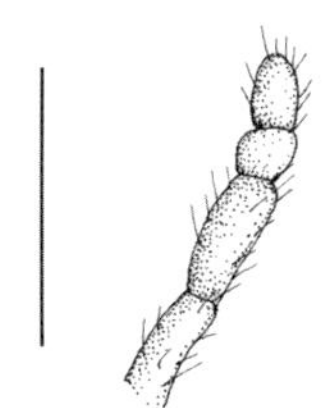

C.3 Distal portion of palp of *Nymphon brevirostre*. Scale bar: 0.5 mm.

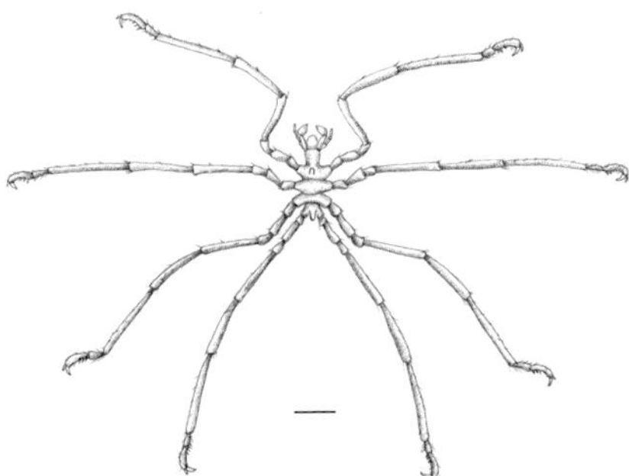

C.4 *Nymphon brevirostre*. Scale bar: 1 mm.

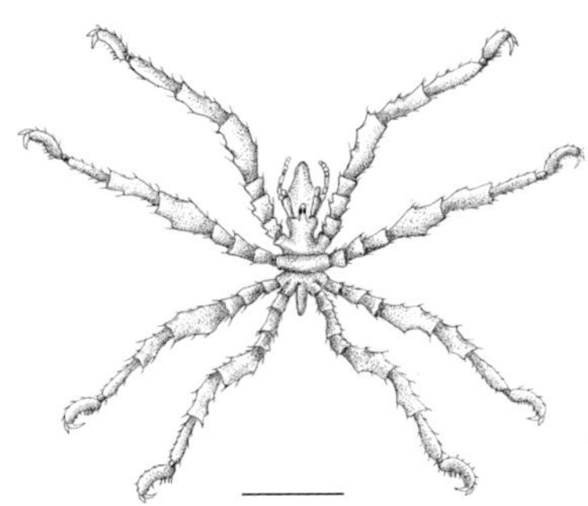

C.5 *Achelia echinata*. Scale bar: 1 mm.

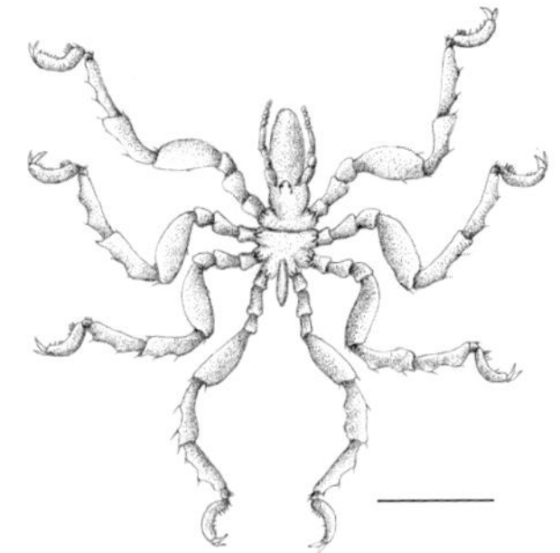

C.6 *Ammothella longipes*. Scale bar: 1 mm.

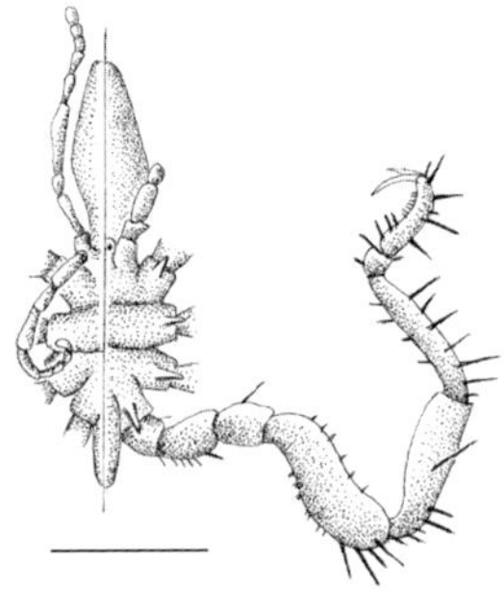

C.7 *Ammothella longioculata*. Scale bar: 1 mm.

bristles on most segments. Lower shore, on hydroids and sponges, amongst a variety of algae (C.3–C.4) *Nymphon brevirostre*

4. Palps with eight segments. Proboscis tapered. Body smooth, up to 2 mm long. Legs with spines and tubercles, up to 6 mm long (C.5). On *Dynamena pumila* and bryozoans *Achelia echinata*

– Palps with nine segments **5**

5. Body smooth, 2 mm long, with only one suture, between segments 1 & 2. Chelifores half length of proboscis. Legs 6 mm long, with few scattered bristles. On *Chondrus*, and other red algae, encrusted with the bryozoans *Flustrellidra hispida* and *Alcyonidium hirsutum* (C.6) *Ammothella longipes*

– Body smooth, 2 mm long, with two sutures, between segments 1 & 2, and 2 & 3. Chelifores more than half length of proboscis. Legs 6 mm long, with numerous spines and bristles. Amongst *Chondrus* and other red algae, encrusted with *F. hispida* and *A. hirsutum*, south-west coasts only (C.7) *Ammothella longioculata*

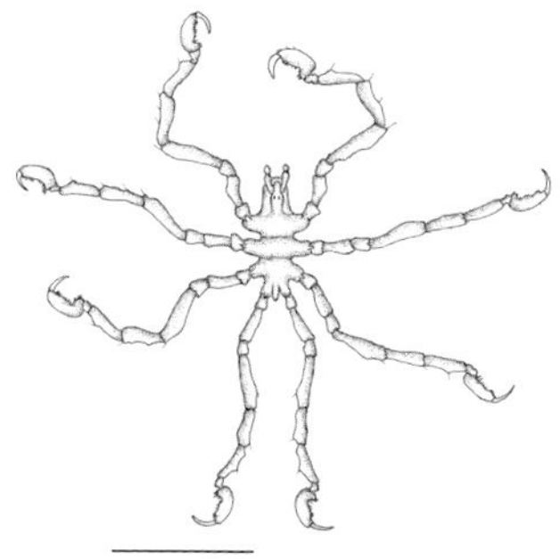

C.8 *Anoplodactylus petiolatus*. Scale bar: 1 mm.

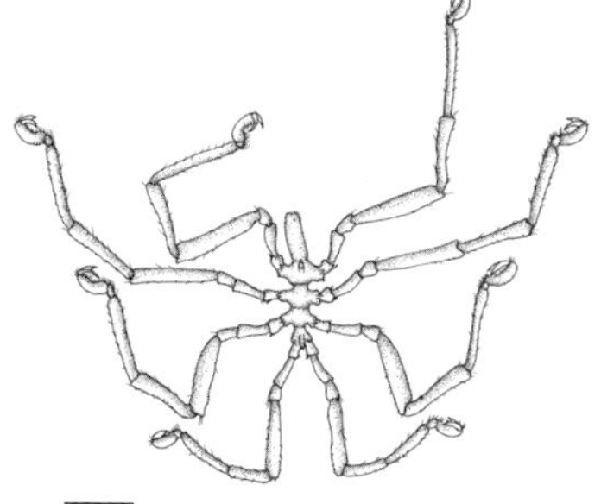

C.9 *Endeis charybdaea*. Scale bar: 2 mm.

6. Chelifores present, palps lacking. Proboscis short, cylindrical; head short and wide with prominent eye tubercle. Body about 1 mm long, legs up to 3 mm. On hydroids and bryozoans (*Amathia* species) (C.8) *Anoplodactylus petiolatus*

– Both chelifores and palps lacking. Body slender, 3–5 mm long; proboscis elongate, half length of trunk. Legs 8–15 mm long, with scattered spines and bristles. Amongst hydroids *Endeis*

Two similar species: *E. spinosa* is widespread, the proboscis is half as long as the trunk and it has few spines; *E. charybdaea* (C.9) is rare on south-west coasts, the proboscis is only one-third the length of the trunk and it bears rows of spines around the mouth. *Phoxichilidium femoratum* resembles *E. spinosa* and occurs in similar habitats. It is distinguished by its smooth legs, lacking spines and bristles, and in possessing chelifores, slightly longer than the proboscis.

Key D Isopods

Isopods display as wide a range of morphology, adaptation and ecological specialisation as any of the large crustacean groups, but in rocky-shore habitats of north-west Europe, while often abundant in terms of numbers of individuals, they display a rather limited taxonomic diversity. No species are especially associated with rock-pool habitats, but the larger species, particularly the species of *Idotea*, are herbivorous and likely to be found in any pool supporting a stable population of their food algae. Some species appear to switch diet as they grow, and juveniles may be commonly found within the red algal fringe. All littoral isopod species recorded from the British Isles are described and illustrated by Naylor & Brandt (2015).

1. Adults with five pairs of thoracic appendages (pereopods). Abdomen (pleon) forming a narrow segmented 'tail'. Male, female and juveniles dissimilar (D.1). A common crevice-dwelling species, often occurring in *Laminaria* holdfasts *Gnathia maxillaris*

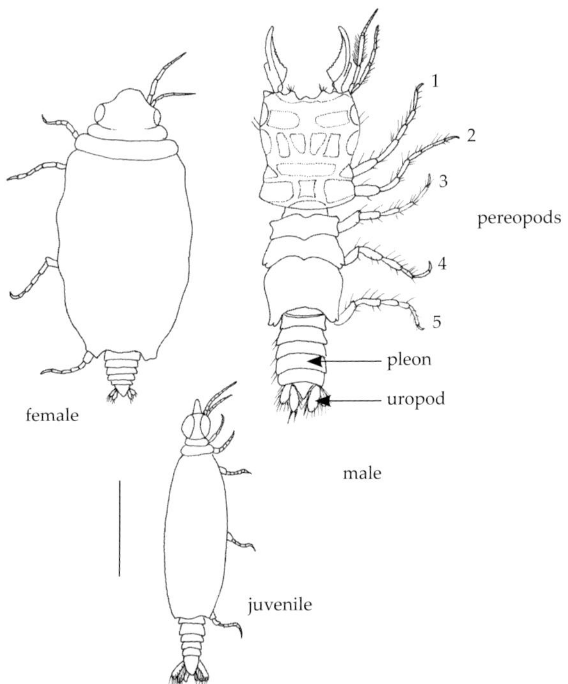

D.1 *Gnathia maxillaris*, female, male and juvenile. Scale bar: 1 mm.

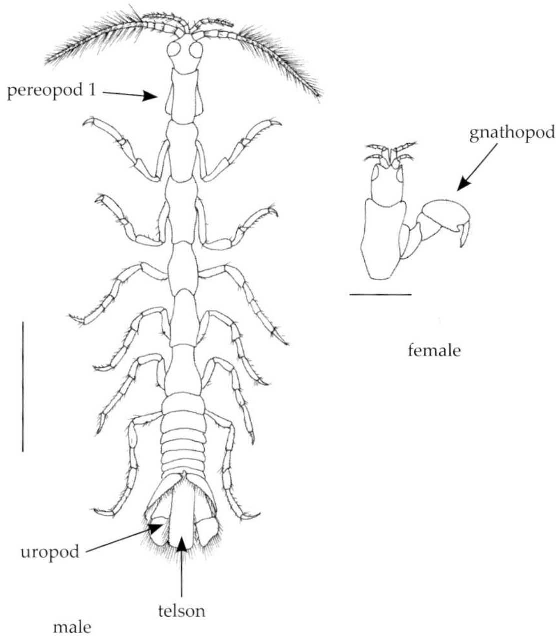

D.2 *Anthura gracilis*, male and anterior end of female. Scale bars: 1 mm.

– Adults with seven pairs of pereopods. Body not as described **2**

2. Uropods attached to the sides of, or beneath, the end portion of the body (D.2, D.6) **3**

– Uropods attached at the very end of the body (D.14–D.15) **12**

3. Uropods at the sides, flattened and forming a tail fan with the telson, or elongate and projecting **4**

– Uropods beneath the body, not visible from above; hinged to form flaps covering the pleopods (D.6) **6**

4. Elongate, almost cylindrical animals with prominent eyes. First pair of pereopods modified to form gnathopods (D.2), often held below body. Uropods biramous, broad, forming a marked tail fan. Females up to 11 mm long, males 4 mm. A crevice-dwelling species, it may occur in kelp holdfasts (D.2) *Anthura gracilis*

– Short, broad animals, strongly convex above, males with paired or single dorsal processes. Uropods biramous or uniramous, narrow and projecting beyond end of body **5**

telson
the posterior, i.e. final, segment of the abdomen (pleon) in arthropods

ramus (pl. rami)
branch. The uropod consists of a basal segment and one or two rami

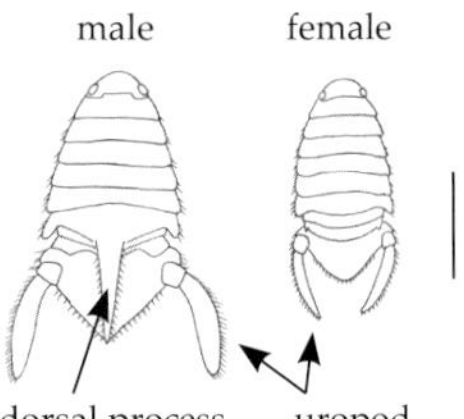

D.3 *Campecopea hirsuta*, male and female. Scale bar: 2 mm.

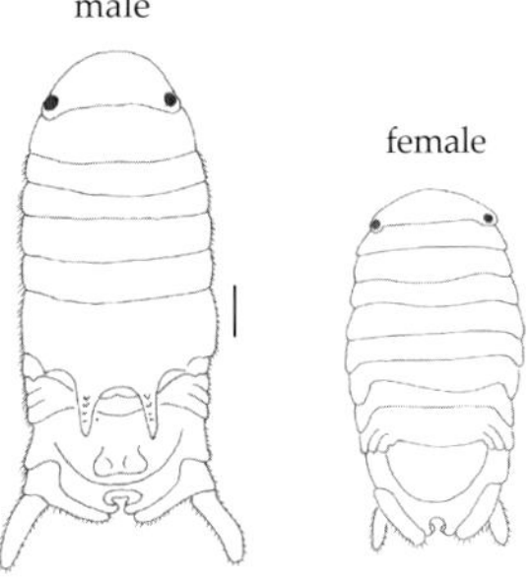

D.4 *Dynamene bidentata*. Scale bar: 1 mm.

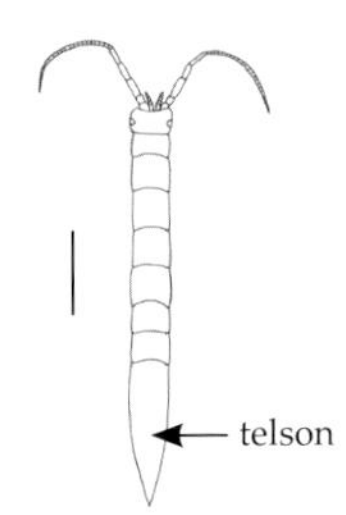

D.5 *Synisoma acuminatum*. Scale bar: 1 cm.

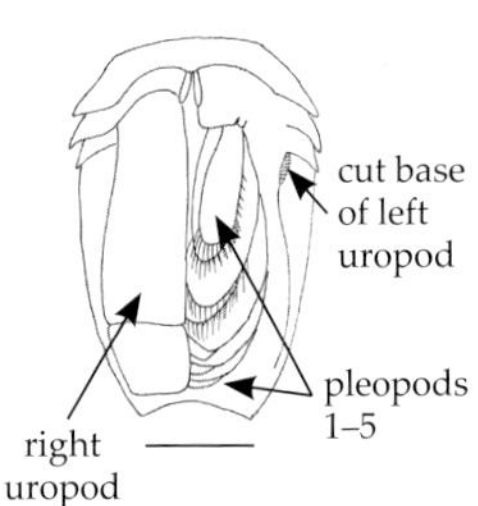

D.6 *Idotea emarginata*, ventral view of telson, to show uropods forming flaps over pleopods. Scale bar: 1 mm.

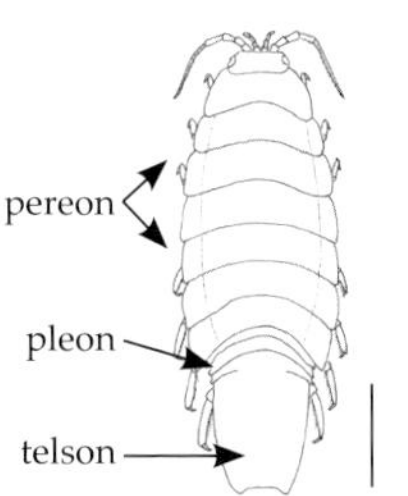

D.7 *Idotea emarginata*, male. Scale bar: 2 mm.

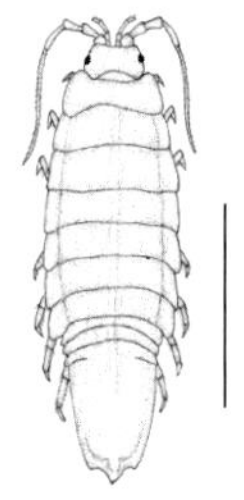

D.8 *Idotea baltica*, male. Scale bar: 5 mm.

5. Uropods uniramous, elongate, projecting conspicuously. Male with a single dorsal process on pereon segment 6. Rolls into a ball when disturbed. Up to 4 mm long. An upper-shore species often common in the lichen *Lichina pygmaea* (D.3) *Campecopea hirsuta*

– Uropods biramous. Male with paired processes on pereon 6 and a blunt tubercle behind. Up to 7 mm long. A crevice-dwelling species often in kelp holdfasts and amongst lower-shore algae (D.4) *Dynamene bidentata*

6. Body with seven distinct segments behind the head. Remaining segments fused with an elongate, terminal telson. Up to 25 mm long; in large, low-shore pools, on *Halidrys siliquosa*, resembling the bladders of the alga (D.5). South and west coasts *Synisoma acuminatum*

– Body with seven pereon segments and two pleon segments (D.7) **7**

7. Posterior margin of telson straight or concave. Males up to 30 mm long, female to 18 mm; light or dark with white markings (D.6–D.7). On detached algae and drift weed, or sometimes amongst bushy fucoids *Idotea emarginata*

– Posterior margin of telson pointed or tapered **8**

8. Posterior margin of telson more or less three-toothed. Males up to 30 mm long, females to 18 mm; green or brown with white spots or lines (D.8). On detached algae and drift weed, often common amongst bushy fucoids *Idotea baltica*

– Posterior margin of telson pointed or rounded, but not three-toothed **9**

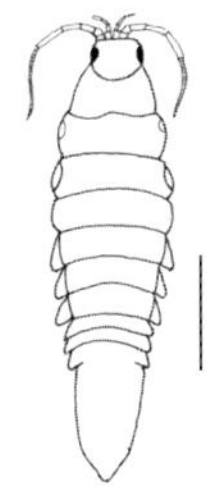

D.9 *Idotea granulosa*, male. Scale bar: 5 mm.

9. Posterior margin of telson with a conspicuous point; slightly convex at the sides. Males up to 20 mm long, females to 13 mm; uniformly red, brown or green, sometimes with longitudinal white marks (D.9). Common in large brown seaweeds on sheltered rocky shores, juveniles in *Cladophora*, *Polysiphonia* and other small algae *Idotea granulosa*

– Telson rounded or with only a blunt point; sides straight or slightly convex **10**

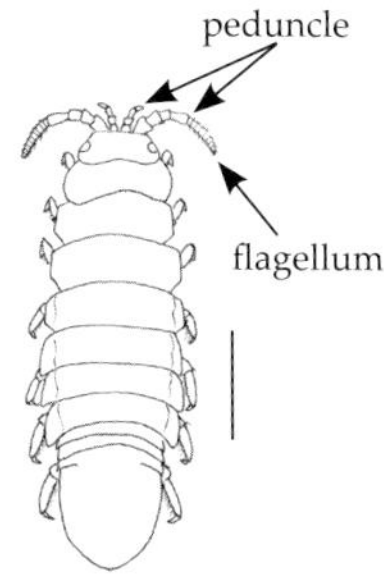

D.10 *Idotea pelagica*, male. Scale bar: 1 mm.

10. The slender flagellum of the antenna much shorter than the basal peduncle, with dense, fine bristles in males. Up to 11 mm long; dark purplish brown with white markings (D.10). Amongst fucoid algae on exposed rocky shores *Idotea pelagica*

– Flagellum of antenna longer than peduncle, without dense bristles **11**

11. Body slender, four to five times as long as wide. Males up to 15 mm long, female to 10 mm; green or brown (D.11). In estuaries amongst lower shore algae or in brackish high-shore pools *Idotea chelipes*

– Body broad, little more than three times as long as wide. Males up to 30 mm long, females to 16 mm; brown, with white streaks or blotches (D.12). On lower shore amongst drift algae, often with *I. emarginata* and *I. baltica* *Idotea neglecta*

12. Body oval, head large with eyes borne on lateral lobes. Antennae as long as body, flagellum shorter than peduncle. Pereopods 2–7 mm long; slender, exceeding body length. A small, delicate animal, up to 3 mm long, appearing rather spider-like (D.13). On kelp holdfasts and amongst erect bryozoans and hydroids on the lower shore *Munna kroyeri*

– Body oval or elongate, eyes on upper surface of head. Antennal flagellum longer than peduncle. Pereopods shorter than body **13**

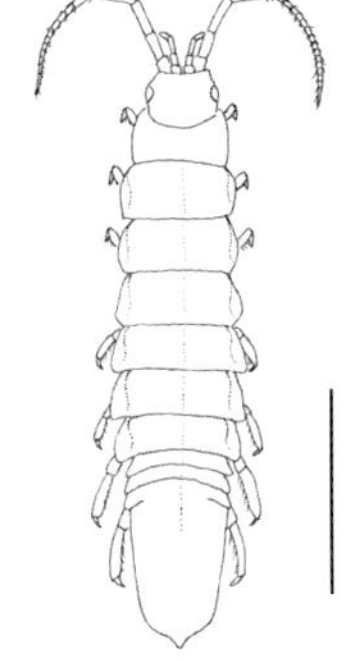

D.11 *Idotea chelipes*, male. Scale bar: 5 mm.

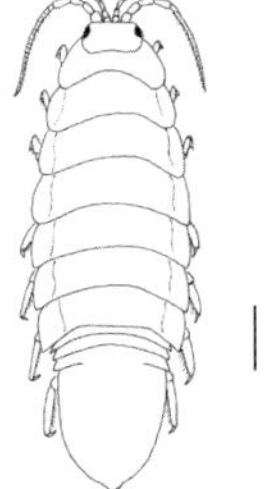

D.12 *Idotea neglecta*, male. Scale bar: 2 mm.

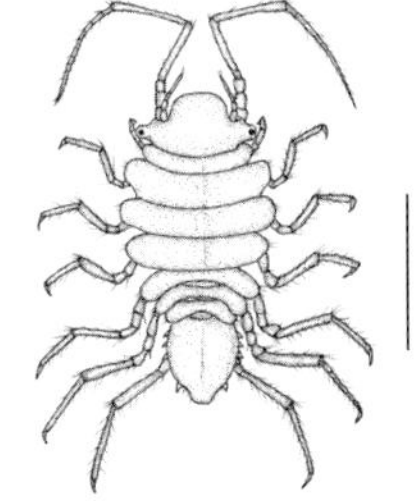

D.13 *Munna kroyeri*, female. Scale bar: 0.5 mm.

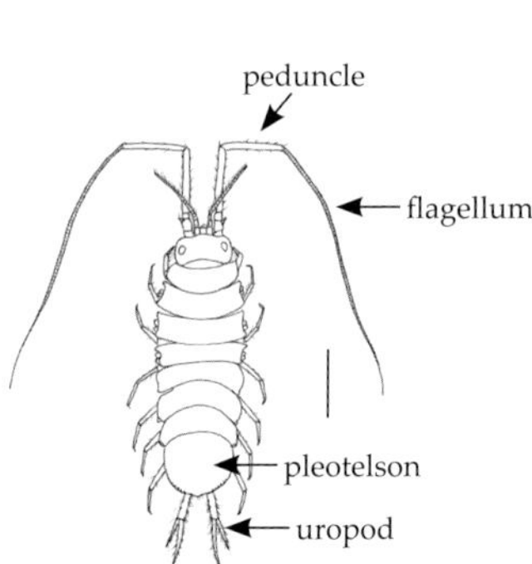

D.14 *Janira maculosa*, male. Scale bar: 1 mm.

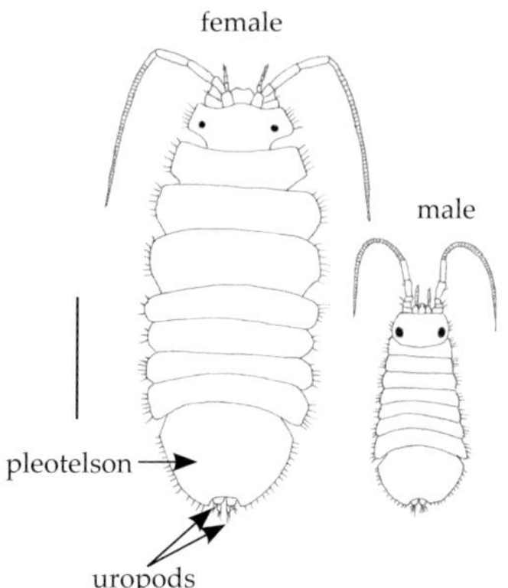

D.15 *Jaera albifrons*. Scale bar: 1 mm.

13. Antennae longer than body. Uropods elongate, longer than pleotelson, projecting conspicuously. Up to 10 mm long (D.14). On kelp holdfasts, on encrusting bryozoans and sponges *Janira maculosa*

– Antennae shorter than body. Uropods tiny, accommodated within a notch in the hind margin of the pleotelson (D.15) *Jaera* species

Key E Amphipods

More than 250 species of the crustacean order Amphipoda are known from coastal and shelf environments of north-west Europe. Identification need not be difficult but usually requires the use of a binocular microscope. The main features of the amphipod body are shown in E.1; the initially intimidating terminology refers to discrete morphological structures which are readily apparent with a minimum of preparation. Characters used in this key are shown and labelled in the accompanying figures. The male usually differs from the female, which is most easily recognised by the presence of eggs beneath the body, or by the plates of the brood chamber between the front series of legs. All British species were described and illustrated by Lincoln (1979), who also provided clear keys for their identification. However, while names of species and genera have changed little, it should be noted that higher-level classification has radically altered in the decades following the publication of Lincoln's monograph. Dichotomous keys for the identification of common representative species of 30 families of amphipod occurring in shallow coastal waters of north-west Europe have been provided by Ashton *et al.* (2017), but Lincoln (1979) remains the only comprehensive account.

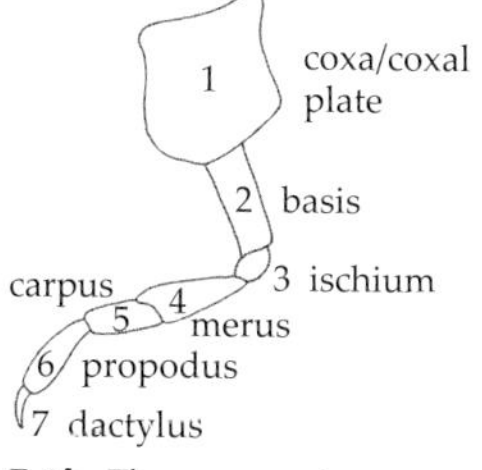

E.1b The segments (articles) of an amphipod limb.

Amphipods are a diverse and often abundant component of the marine, shallow water benthos, occupying a range of habitats. On rocky shores they occur in silty microhabitats beneath stones and in algal holdfasts, either free-living or constructing permanent tubes; some species are associated

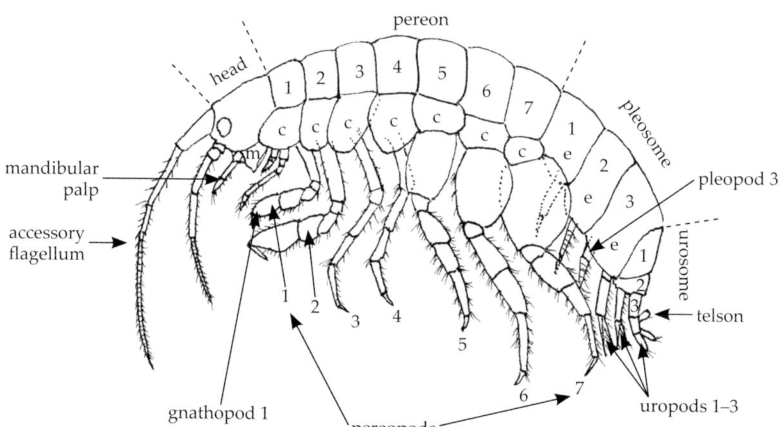

E.1a Main features of amphipod morphology: m = mandible, c = coxal plate, e = epimeral plate.

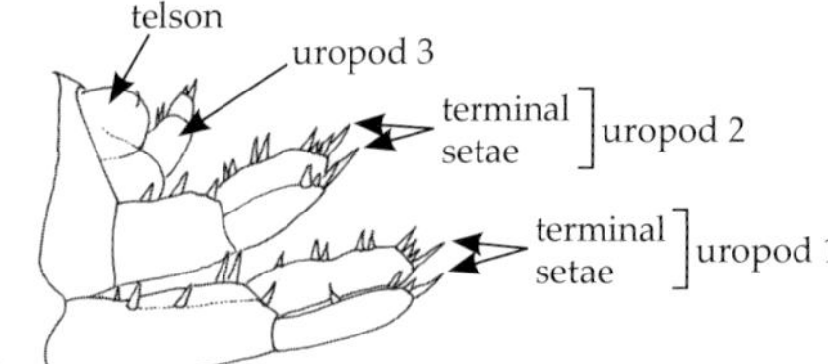

E.2 *Hyale nilssoni*, male, uropods: note that the paired rami of uropods 1 & 2 each terminate in an especially stout seta. (After Lincoln 1979.)

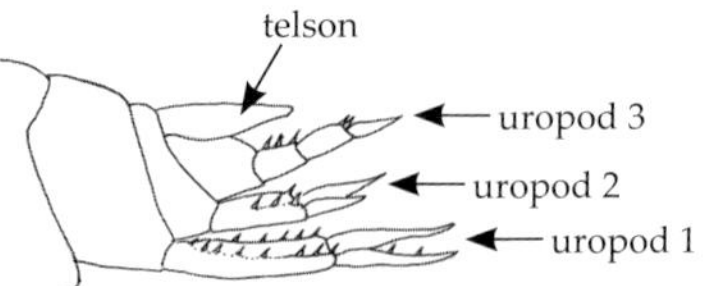

E.3 *Stenothoe monoculoides*, female, uropods: note that the paired rami of uropods 1 & 2 do not bear stout terminal setae. (After Lincoln 1979.)

with particular species of sedentary invertebrates, as commensals or parasites. No species of amphipod is known to be characteristic of rock-pool habitats, but as associates of sponges, hydroids, bryozoans and other sessile animals, and as part of the algal epiphyte community, amphipods are likely to occur in any pool providing appropriate ecological conditions. The following key should facilitate identification of those species – and **only those species** – most commonly associated with seaweeds.

1. Body elongate, slender, consisting of seven pereon segments, the first fused with the head. Typically with two paired appendages anteriorly and two or three pairs posteriorly, with or without reduced appendages in between (A.8) **41**

– Not as described. Body clearly regionated: head, pereon with seven paired appendages, pleon with three and urosome with two or three, all distinct (E.1a) **2**

2. Rami of uropods 1 and 2 each with a stout, robust seta at the tip, sometimes partly hidden by clumps of smaller bristles, but always present (E.2) **16**

– Rami of uropods 1 and 2 with or without bristles and/or setae along their length but without a stout, robust seta at the tip (E.3) **3**

3. Antenna 1 very short, peduncle with large, broad basal segment, and short second and third segments (E.4–E.6); flagellum scarcely longer than peduncle, accessory flagellum conspicuous. Gnathopod 2 long and slender, with minute dactylus **4**

– Not with the above association of characters **6**

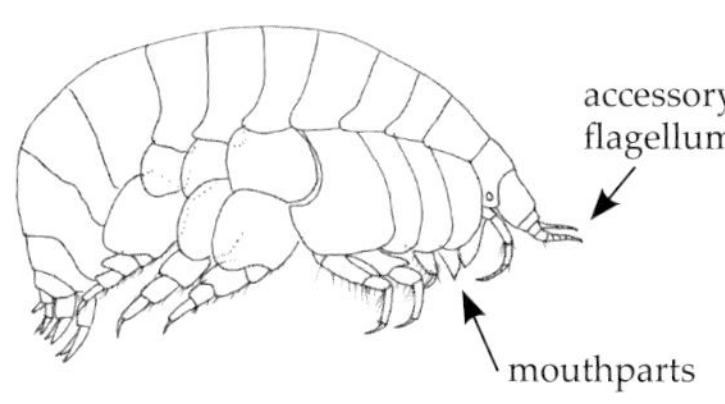

E.4 *Acidostoma sarsi*, female.

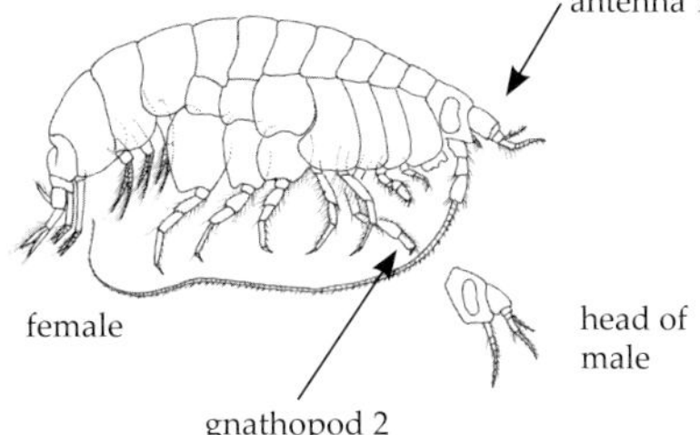

E.5 *Orchomene humilis*.

4. Head with piercing mouthparts, formed into conical bundle; eyes small and round. Up to 9 mm long; pinkish red (E.4). In kelp holdfasts *Acidostoma sarsi*

– Mouthparts not as described, eyes large and oval **5**

5. Telson with a small split extending about one-third of its length. Gnathopod 2 with long, slender propodus. Up to 8 mm long; pale yellow or cream, sometimes with red flecks (E.5). In kelp holdfasts *Orchomene humilis*

– Telson with elongate split extending more than half its length. Gnathopod 2 with short, fat propodus. Up to 5 mm long; greyish or white (E.6). Common on all coasts; often in kelp holdfasts *Tryphosa nana*

6. First coxal plate small, partly or completely hidden by second; plates 2–4 very large (E.9–E.12). No accessory flagellum. Telson undivided **7**

– Not with the above association of characters **10**

7. Uropod 3 biramous, each ramus with one segment only **8**

– Uropod 3 uniramous, the ramus with two segments **9**

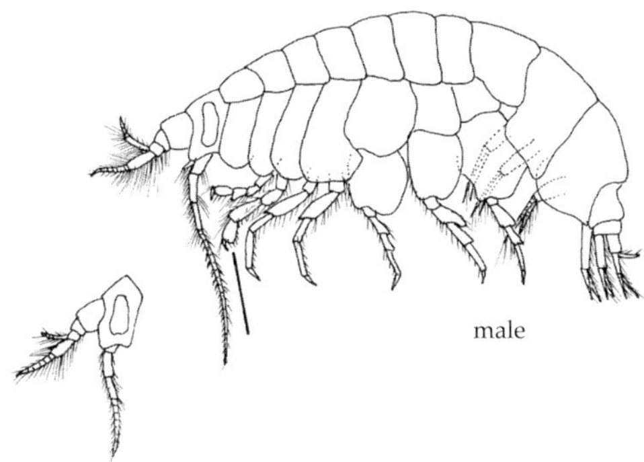

E.6 *Tryphosa nana*.

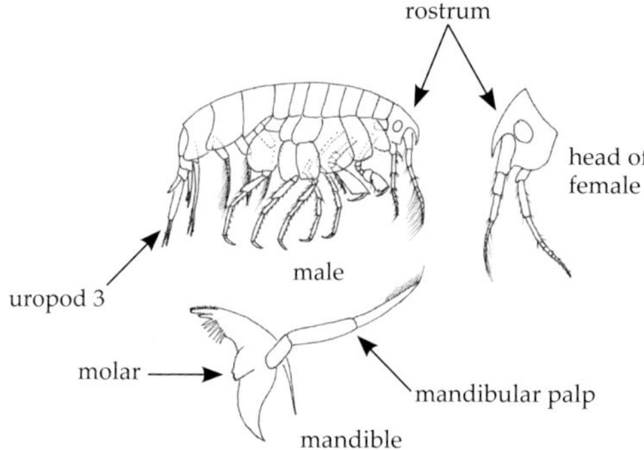

E.7 *Amphilochus manudens.*

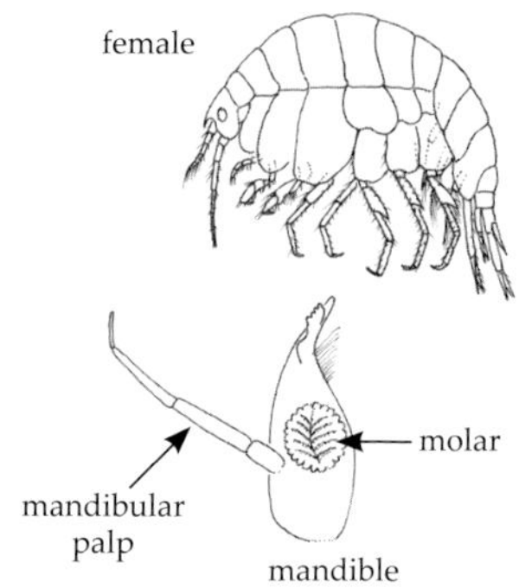

E.8 *Gitana sarsi.*

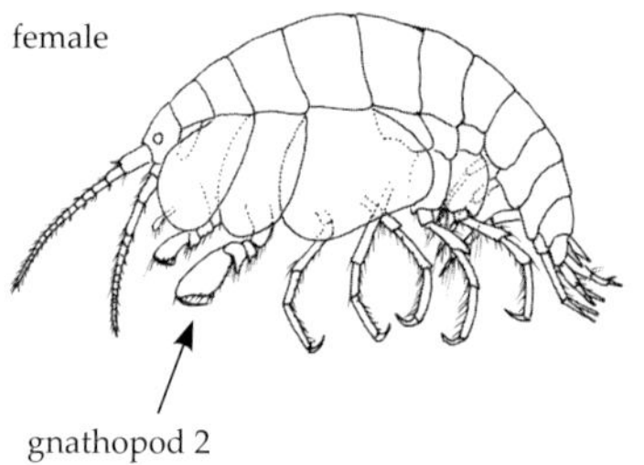

E.9 *Stenothoe monoculoides*, female.

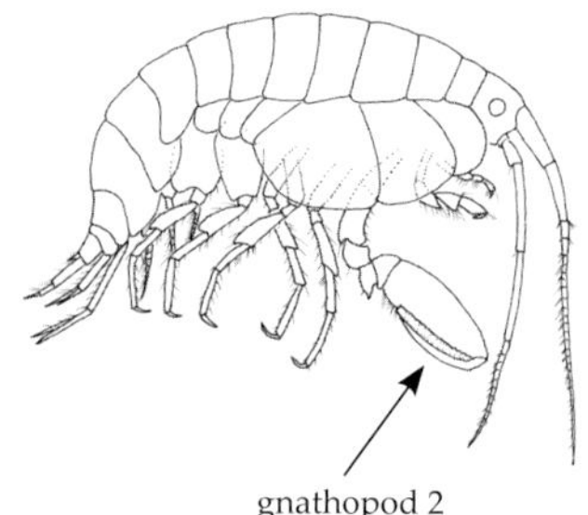

E.10 *Stenothoe marina*, female.

8. Head with large, curved rostrum; antennae short, of equal length. Mandible with small molar. Up to 5 mm long; pale brown or red (E.7). Lower shore, in kelp holdfasts *Amphilochus manudens*

– Head with short, curved rostrum, antenna 2 longer than 1. Mandible with prominent ridged molar. Up to 3 mm long; broadly banded with dark brown or black/violet (E.8). All coasts, common. In kelp holdfasts and amongst fucoids *Gitana sarsi*

9. Gnathopod 2 with rectangular propodus, palm (surface facing dactylus) almost perpendicular to long axis. Eyes small, round, red. Up to 3 mm long; white with red flecks (E.9). All coasts, common. In kelp holdfasts and amongst lower shore fucoids *Stenothoe monoculoides*

– Gnathopod 2 with elongate oval propodus, palm almost parallel to long axis. Eyes large, round, dark red. Up to 6 mm long; white with yellow and pink markings (E.10). All coasts, common. In kelp holdfasts *Stenothoe marina*

10. Gnathopod 1 carpochelate (i.e. dactylus forms pincer with carpus); gnathopod 2 much larger than 1, massive.

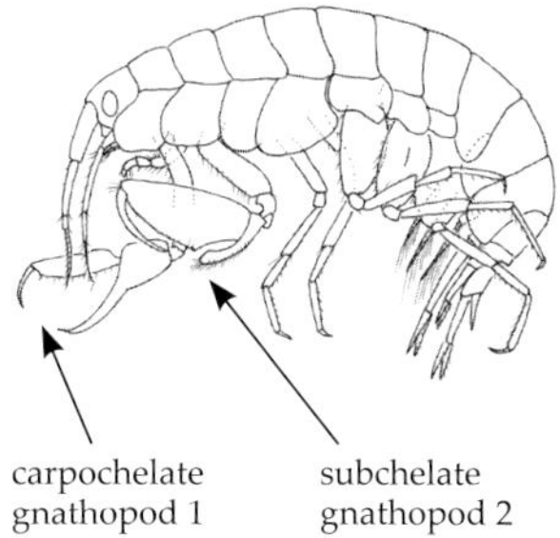

E.11 *Leucothoe spinicarpa*, female.

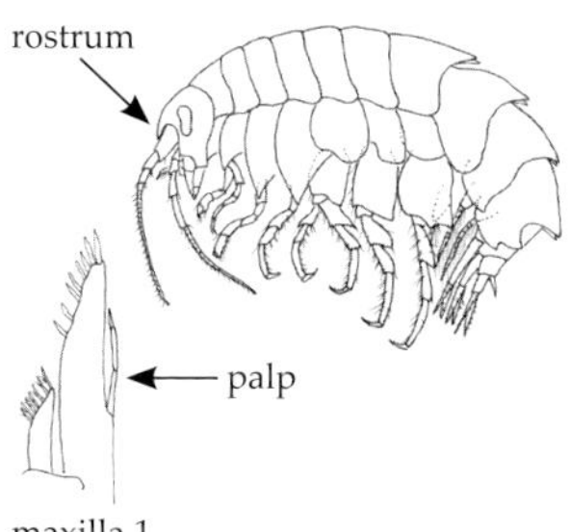

E.12 *Iphimedia minuta*, female.

Eyes large, round, red. Up to 18 mm long; pinkish with dark bands, sometimes pale green above (E.11). All coasts, often common. Lower shore, amongst fucoid algae, in kelp holdfasts, often associated with sponges and ascidians *Leucothoe spinicarpa*

(*L. incisa* is smaller than *L. spinicarpa* – up to 7 mm – and differs also in having a much smaller dactyl to gnathopod 1, less than one-quarter length of propodus.)

– Not with the above association of characters **11**

11. Head with large curved rostrum and prominent eyes. Piercing mouthparts forming a conical bundle; mandible slender (E.12) **12**

– Not with the above association of characters **13**

12. Maxilla 1 with very small palp (use fine forceps to pull out the mouthparts). Up to 6 mm long; yellowish with dark red or brown bands (E.12). Widespread and common. Lower shore, amongst algae, often in kelp holdfasts *Iphimedia minuta*

– Maxilla 1 with well-developed palp. Up to 12 mm long; white or yellowish, with red or pink bands (E.13). All coasts. Lower shore, often in kelp holdfasts *Iphimedia obesa*

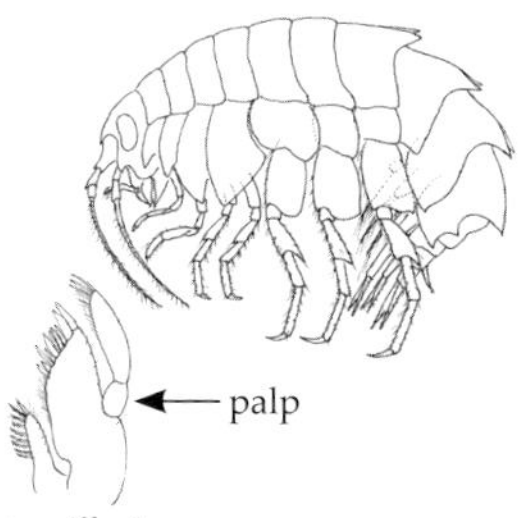

E.13 *Iphimedia obesa*, female.

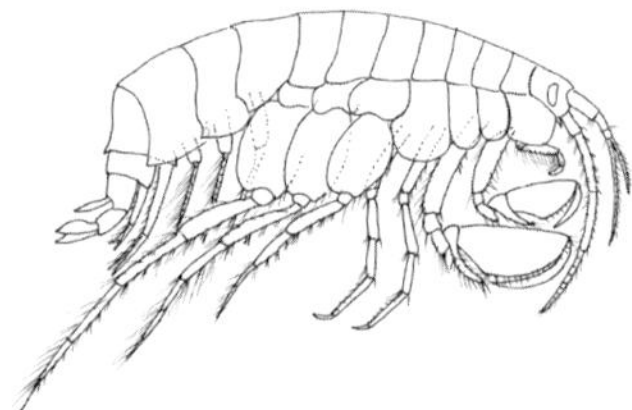

E.14 *Liljeborgia pallida*, male.

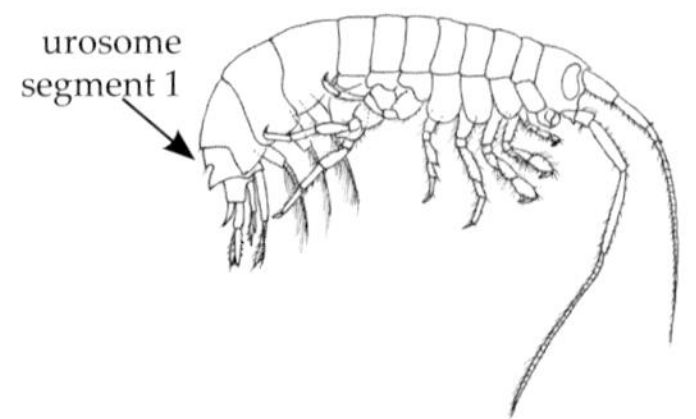

E.15 *Nototropis swammerdami*, male.

13. Urosome with three distinct segments. Coxal plates large, overlapping. Gnathopods 1 and 2 large, subchelate (i.e. dactylus folded back along lower edge of propodus); pereopods 3 and 4 short and very slender, 5 to 7 longer, each with broad basis. Antenna 1 shorter than 2; accessory flagellum about half length of antenna 1 flagellum. Eyes large, oval. Up to 10 mm long; pale orange, or white with red or orange patch (E.14). West coasts only. Lower shore, in kelp holdfasts *Liljeborgia pallida*

– Urosome segments 2 and 3 fused. Coxal plates not as described. Gnathopods 1 and 2 small. Antenna 1 and 2 almost equal length, or 1 slightly shorter than 2, no accessory flagellum. Pereopods 3 and 4 not more slender than 5 to 7 **14**

14. Head with conspicuous rostrum and large eyes. Mandible with a palp, visible between or below bases of antennae (E.15). Urosome segment 1 with a small dorsal tooth and a larger posterior dorsal tooth. Telson cleft. Up to 10 mm long; white, with brown patches. Widespread and common. Lower shore, amongst kelp holdfasts *Nototropis swammerdami*

– Rostrum not evident. Mandible without a palp. Pleon smooth, or with large dorsal processes; urosome segment 1 with at most one small dorsal tooth **15**

15. Pereopods 3–7 with merus longer than combined length of carpus and propodus. Urosome segment 1 with single small dorsal tooth posteriorly. Up to 6 mm long; white with brown patches (E.16). Widespread and common.

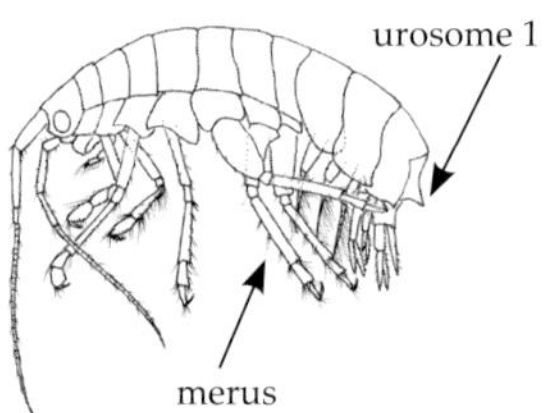

E.16 *Tritaeta gibbosa*, female.

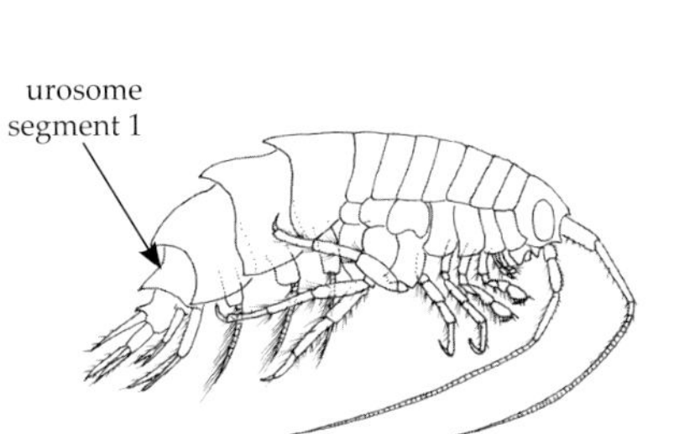

E.17 *Dexamine spinosa*, male.

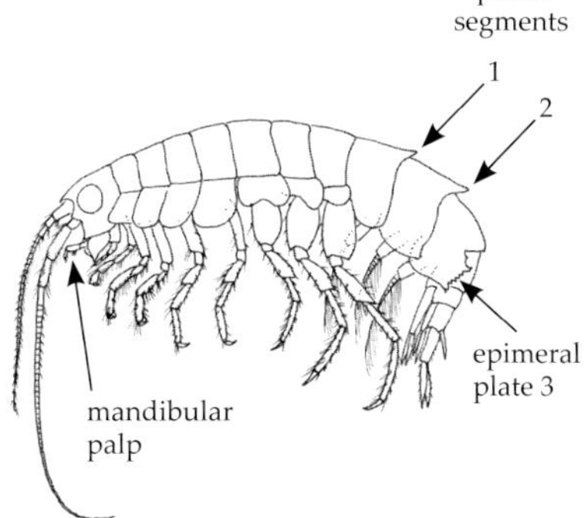

E.18 *Apherusa bispinosa*, female.

Lower shore, particularly associated with sponges and ascidians, often in kelp holdfasts *Tritaeta gibbosa*

– Pereopods 3–7 with merus shorter than combined length of carpus and propodus. Pleon segments each with prominent tooth-like process dorsally. Up to 14 mm long, boldly coloured: bright red or red-brown, often with white spots and black patches (E.17). Widespread and common. Lower shore, in kelp holdfasts *Dexamine spinosa*

16. Mandible with a palp, visible between or below bases of antennae **17**

– Mandible without a palp **39**

17. Telson elongate, flat, undivided or with a longitudinal split medially. Coxal plate 4 may have concave posterior margin **18**

– Telson short and thick, often fleshy; undivided or at the most with a small notch at tip **24**

18. Antenna 1 without an accessory flagellum **19**

– Antenna 1 with a conspicuous accessory flagellum (E.21) **20**

19. Dorsal processes on pleon segments 1 and 2. Head with prominent rostrum and large round eyes. Epimeral plate 3 strongly toothed on hind margin. Up to 7 mm long; white to purplish, often with dark blotches (E.18). Widespread and common. Lower shore, in kelp holdfasts and *Corallina* turf *Apherusa bispinosa*

– No processes. Head with large, kidney-shaped eyes. Up to 8 mm long; yellowish white, pink, or pinkish purple with white dorsal patch, with bright red-brown spots

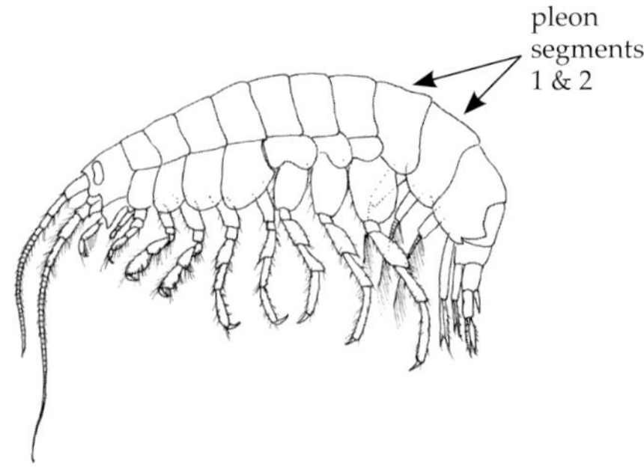

E.19 *Apherusa jurinei*, female.

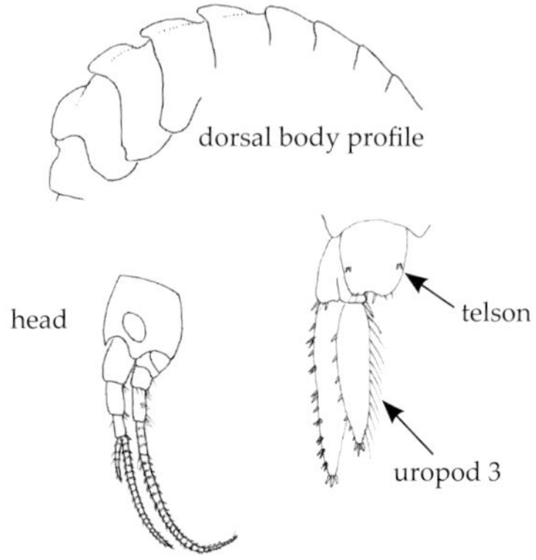

E.20 *Gammarellus angulosus.*

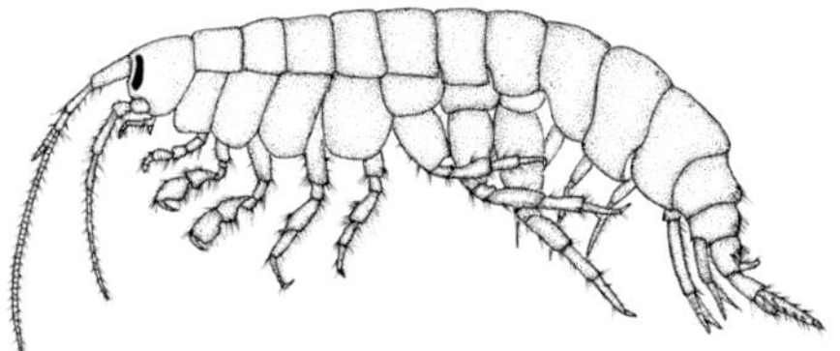

E.21 *Echinogammarus marinus.*

or blotches (E.19). All coasts, common. Lower shore, amongst algae and in kelp holdfasts *Apherusa jurinei*

20. Body with conspicuous, sharp, dorsal ridge. Antennae 1 and 2 of equal length. Eyes large. Gnathopods 1 and 2 about equal size. Up to 15 mm long; yellowish with red-brown blotches (E.20). Widespread and common. Intertidal and sublittoral, in kelp holdfasts and red algal turfs *Gammarellus angulosus*

    (*G. homari* is a larger species – up to 35 mm long – with small eyes, and the dorsal ridge strongly extended posteriorly.)

– Not with the above association of characters **21**

21. Gnathopods 1 and 2 subchelate, about equal size. Dorsal surface of urosome with groups of short spines. Inner ramus of uropod 3 very small. Eyes very large, kidney shaped. Up to 25 mm long; dark green, sometimes with red or yellow patches (E.21). Widespread and common. Intertidal, associated with a wide range of algae *Echinogammarus marinus*

– Gnathopod 2 much larger than 1. Urosome spines sparse or absent **22**

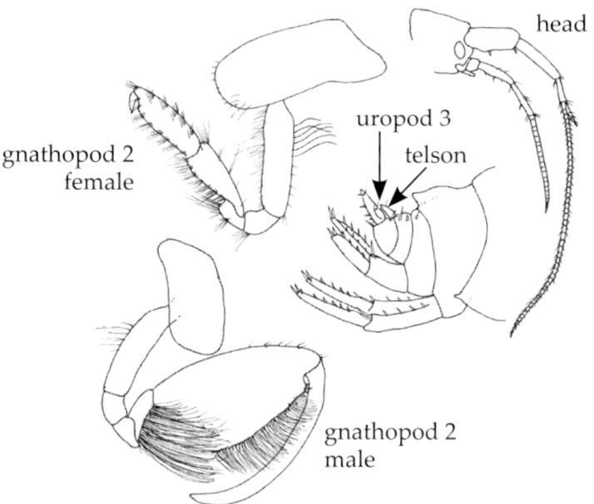

E.22 *Gammarella fucicola.*

22. Outer ramus of uropod 3 very short. Urosome with dorsal ridge. Eyes small and round. Up to 10 mm long; yellow-brown (E.22). South-west and west coasts. Intertidal and sublittoral, amongst fucoids and kelp holdfasts *Gammarella fucicola*

– Uropod 3 rami both of about equal length **23**

23. Uropod 3 rami short and fat, about as long as peduncle. Telson deeply cleft; each lobe rounded at the tip, with small groups of spines just in front of tip. Up to 10 mm long; yellow-white flecked with violet, or violet with white patches (E.23). South-west and west coasts. Lower shore, amongst fucoids and in kelp holdfasts *Elasmopus rapax*

– Uropod 3 rami very elongate. Telson widely cleft, with spines at tip of each lobe. Antenna 1 shorter than 2. Gnathopods 1 and 2 simple in female; gnathopod 1 simple, gnathopod 2 subchelate in male. Up to 10 mm long; yellow or orange, with red blotches (E.24). All coasts. Lower shore, in kelp holdfasts *Cheirocratus sundevallii*

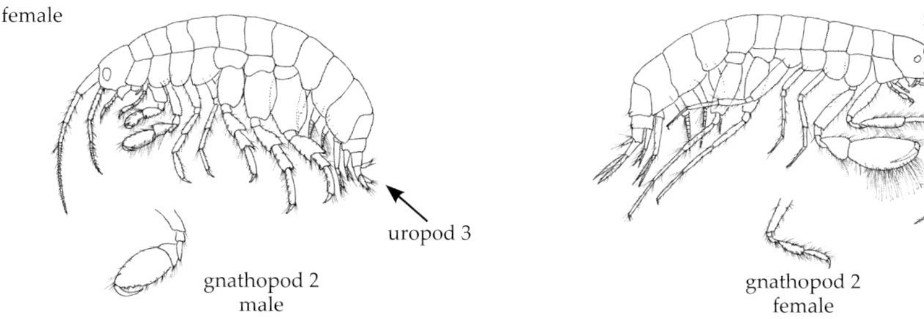

E.23 *Elasmopus rapax.*

E.24 *Cheirocratus sundevallii.*

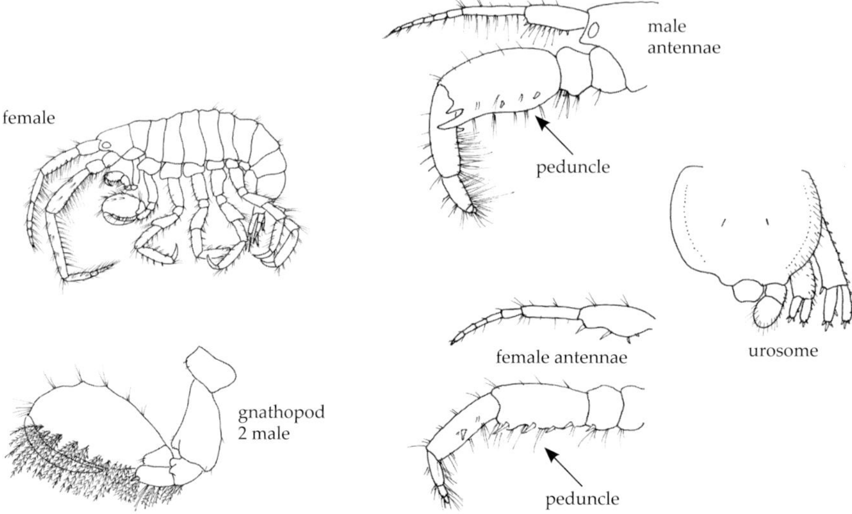

E.25 *Podocerus variegatus.*

E.26 *Apocorophium acutum.*

24. Uropods 1 and 2 distinct, spiny; uropod 3 reduced to inconspicuous plate beside telson. Urosome segment 1 more than twice length of segment 2. Up to 4 mm long; dark red or brown-red, sometimes with purple patch (E.25). South-west and west coasts. Intertidal and shallow subtidal, in kelp holdfasts and *Corallina* turf *Podocerus variegatus*

– Urosome with three pairs of uropods **25**

25. Antenna 2 longer than 1, with massively developed peduncle (E.26). Urosome flattened. Uropods 1 and 2 biramous; uropod 3 short, uniramous **26**

– Not with the above association of characters **28**

26. Urosome segments fused; viewed from above, distinct lateral ridges hide insertion of uropods. Antenna 1 of female with four large teeth on lower edge of segment 1; antenna 2 of male with one large and one small tooth on tip of lower edge of segment 4 and a single tooth near base of segment 5. Up to 4 mm long (E.26). South-west only; builds tubes amongst algae, hydroids and sponges. Locally abundant *Apocorophium acutum*

– Segments fused, but no lateral ridges; uropods inserted in lateral notches, visible from above **27**

27. Female: antenna 2 with few, unpaired spines on lower surface of segment 4, no spines on upper edge; uropod 1 peduncle without bristles on outer margin. Male: antenna 1 with seven spines on lower edge of segment

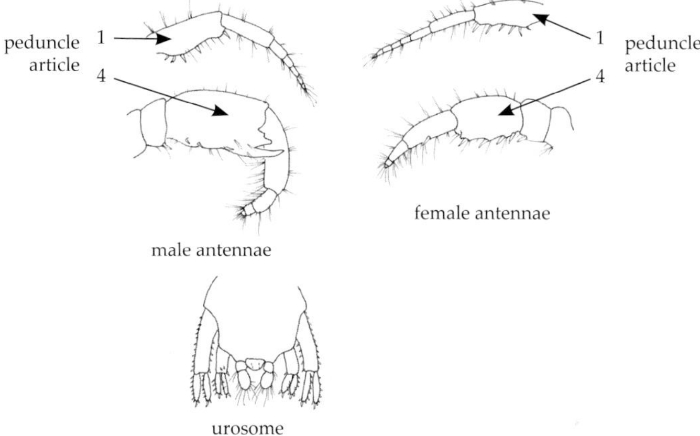

E.27 *Monocorophium sextonae.*

1; uropod 1 peduncle with spines along whole outer margin. Up to 5 mm long (E.27). South-west only, lower shore. Builds tubes in kelp holdfasts, hydroids and sponges *Monocorophium sextonae*

– Antenna 2 with 1–2 pairs of spines on lower surface of segment 4; antenna 1 with tiny curved spine near base of segment 1. Uropod 1 peduncle with 3–4 spines on inner margin. Up to 5.5 mm long (E.28). West and south-west coasts, lower shore and shallow subtidal. Builds tubes in *Corallina* turf, kelp holdfasts, amongst algae and hydroids *Crassicorophium bonellii*

28. Uropod 3 rami shorter than peduncle (E.29) **29**

– Uropod 3 rami as long as or longer than peduncle **35**

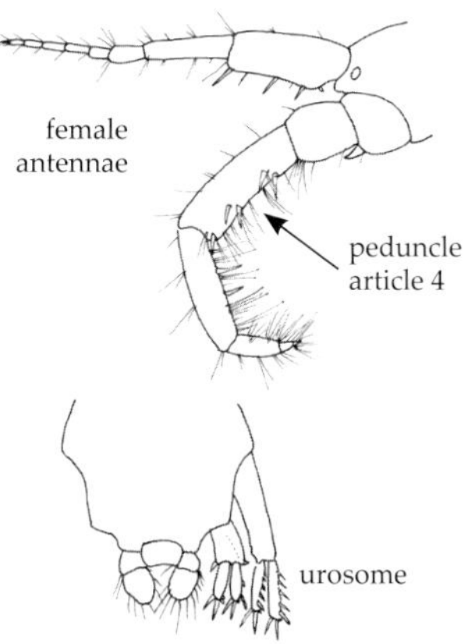

E.28 *Crassicorophium bonellii.*

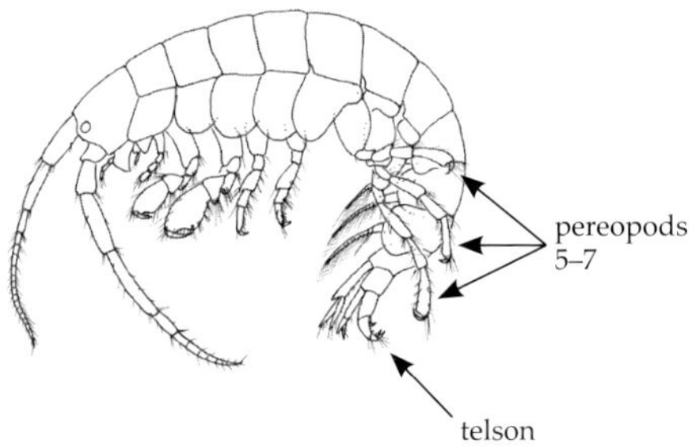

E.29 *Ampithoe gammaroides*, female.

29. Article 3 of antenna 1 less than half length of article 2 (E.29) **30**
– Article 3 of antenna 1 more than half length of segment 2 **31**

30. Pereopods 5–7 with propodus broadened at the end, forming subchelate closure with dactylus (E.29). Telson with pair of large, curved spines on posterior edge. Antennae half length of body, with sparsely distributed bristles. Up to 8 mm long. South-west and west coasts. Lower shore, amongst algae and in kelp holdfasts *Ampithoe gammaroides*
– Propodus of pereopods 5–7 only slightly expanded at the end, not clearly subchelate. Telson without spines on posterior edge. Eyes small. Up to 20 mm long; red to green with dark spots, often with white patches (E.30). Widespread and common. Builds tubes amongst wide variety of algae, in red algal turfs and in holdfasts *Ampithoe rubricata*

31. Uropod 3 uniramous. Body markedly flattened dorso-ventrally. Head elongate, with very large eyes. Up to 10 mm long; brown to orange with darker patches. Builds tubes of fine sand and silt, in *Zostera* beds, amongst algae, in algal turfs, hydroids and kelp holdfasts, lower shore and, sublittoral. Recorded from most coasts in the

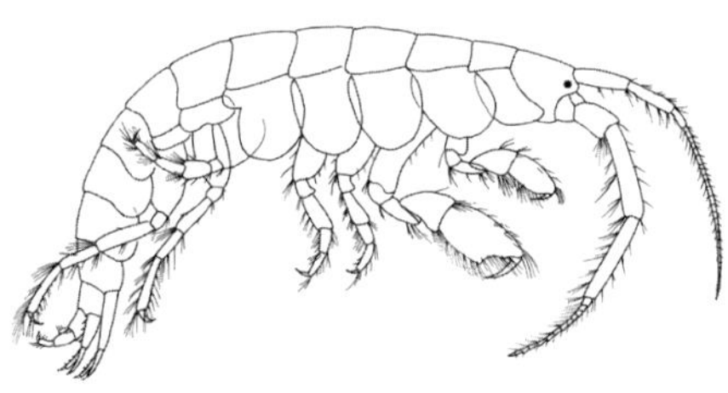

E.30 *Ampithoe rubricata*, male.

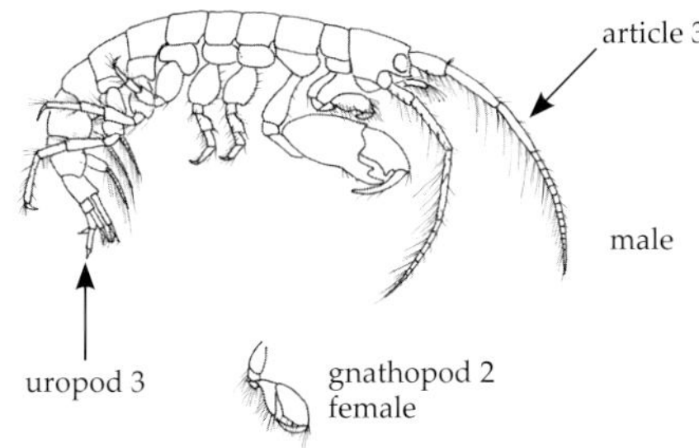

E.31 *Ericthonius brasiliensis.*

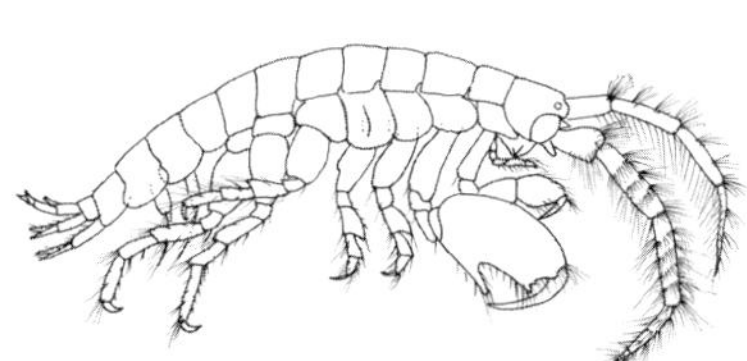

E.32 *Parajassa pelagica*, male.

region, but apparently absent or scarce on south-west coasts of North Sea. Two similar species, distinguished principally by the carpus of gnathopod 2, which bears two processes in *Erichthonius brasiliensis* (E.31) but just one in *E. difformis* *Erichthonius brasiliensis* and *Ericthonius difformis*

– Uropod 3 biramous **32**

32. Accessory flagellum indistinct, reduced to minute tubercle. Antenna 2 very robust, with thick girdles of bristles. Up to 9 mm long; grey with brown bands (E.32). Widespread and common. Lower shore, in holdfasts, amongst hydroids and in algal turfs *Parajassa pelagica*

– Accessory flagellum distinct, though short **33**

33. Coxal plate 1 very small, mostly concealed by plate 2; plates 2–4 broad and elongate. Antennae with long, sparse bristles. Eyes large. Up to 3 mm long; brown (E.33). West coasts. Lower shore, in kelp holdfasts *Microjassa cumbrensis*

– Coxal plate 1 distinct; plates 1–4 of similar size, plate 6 much smaller than plate 5 **34**

34. Gnathopod 2 of male with single large 'thumb' on palm of propodus. Body slender, rather flattened. Antennae stout, with dense bristles. Eyes small. Up to 12 mm long; white with characteristic red-brown spots, blotches and

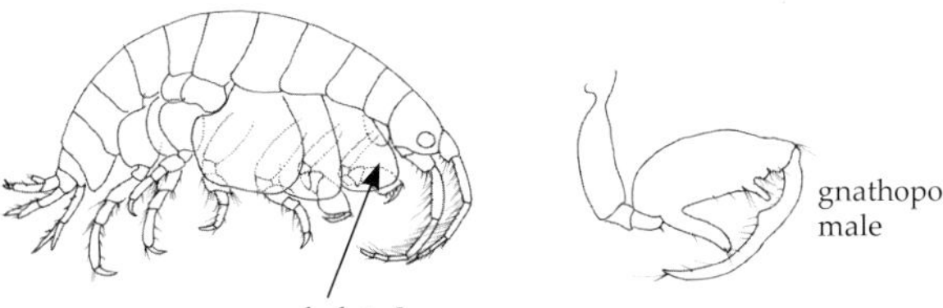

E.33 *Microjassa cumbrensis.*

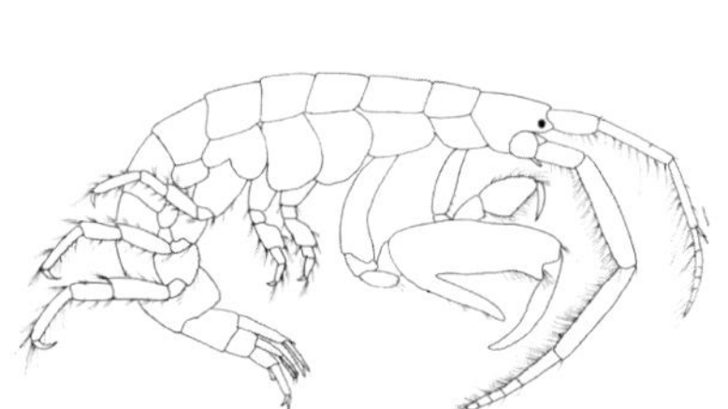

E.34 *Jassa falcata*, male.

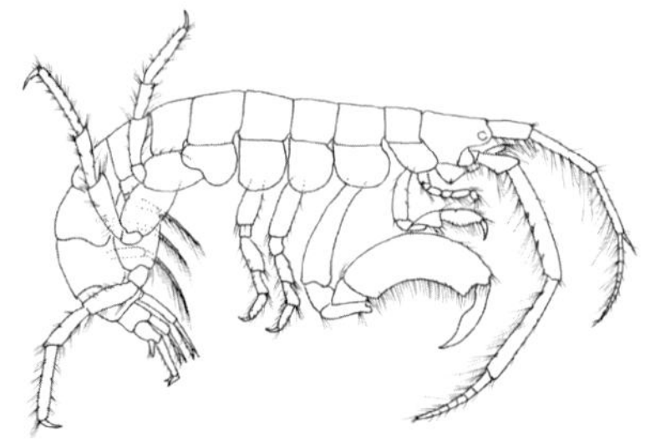

E.35 *Ischyrocerus anguipes*, male.

streaks (E.34). Widespread and common. Builds tubes in algal turfs, amongst hydroids and in holdfasts *Jassa falcata*

– Gnathopod 2 of male with concave, densely hairy palm to propodus, lacking a 'thumb'. Eyes relatively large. Up to 10 mm long, sometimes larger; light yellow green with dark spots, or white with brown bands (E.35). Widespread and common. Lower shore, builds tubes amongst algae and in kelp holdfasts *Ischyrocerus anguipes*

Note that females of *Jassa* and *Ischyrocerus* species are difficult to distinguish.

35. Gnathopod 1 larger than gnathopod 2 (conspicuously in male, less so in female) **36**

– Gnathopod 1 smaller than gnathopod 2 in both sexes **38**

36. Gnathopod 1 of male with simple carpus; dactyl opposed by a short tooth on lower margin of the propodus. Both gnathopods of male with long, dense bristles on upper surface. Up to 6 mm long; white with brown bands (E.36). Widespread and common. Intertidal and shallow sublittoral, amongst algae and in kelp holdfasts *Lembos websteri*

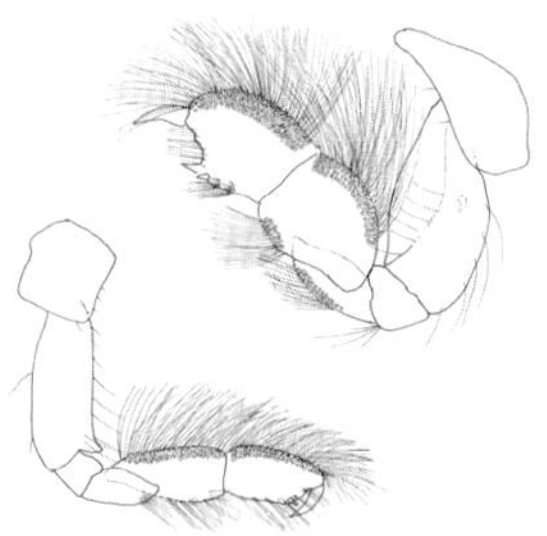

E.36 *Lembos websteri*, male, gnathopods 1 & 2.

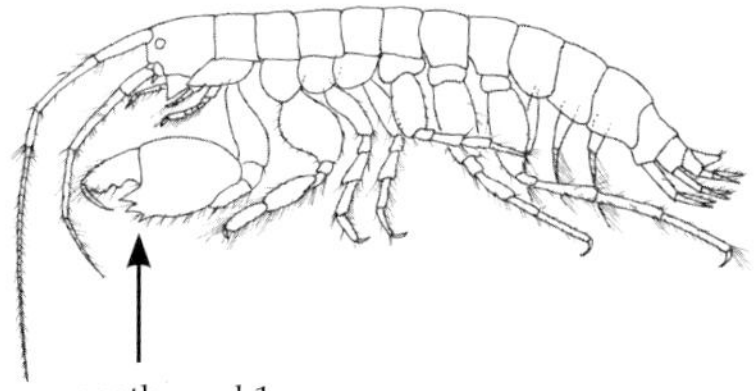

E.37 *Microdeutopus gryllotalpa*, male.

– Gnathopod 1 of male with small propodus; dactyl opposed by one or more projecting teeth on lower margin of carpus (E.37) **37**

Note that females of *Lembos websteri* and the *Microdeutopus* species in couplet 37 cannot be easily distinguished.

37. Gnathopod 1 of male with two to four distinct teeth on lower edge of carpus; gnathopod 2 of male without comb of bristles. Up to 10 mm long (E.37). Widespread and common. Intertidal, amongst algae and in holdfasts *Microdeutopus gryllotalpa*

– Gnathopod 1 of male with a single large tooth on lower margin of carpus; gnathopod 2 of male with comb of coarse bristles on upper surface. Up to 8 mm long (E.38). West coasts. Lower shore, in algal turfs and kelp holdfasts *Microdeutopus versiculatus*

Note that females of these two species and *Lembos websteri* (couplet 36) cannot be easily distinguished.

38. Uropod 3 uniramous. Coxal plates large, with fine hairs along the edges. Antennae only one-third body length. Eyes small and round. Up to 3 mm long; darkly pigmented (E.39). South and west coasts in kelp holdfasts *Microprotopus maculatus*

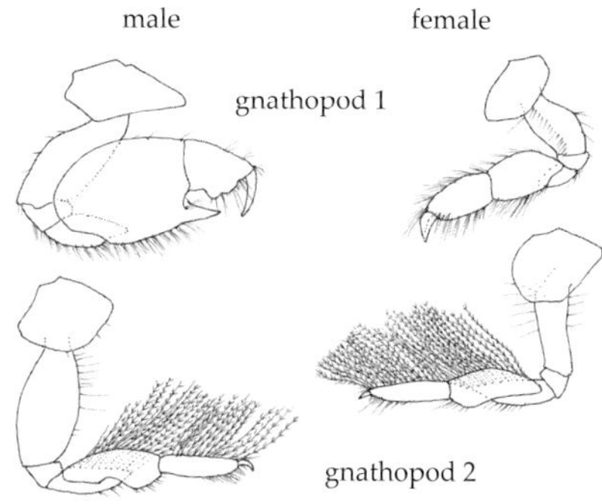

E.38 *Microdeutopus versiculatus*.

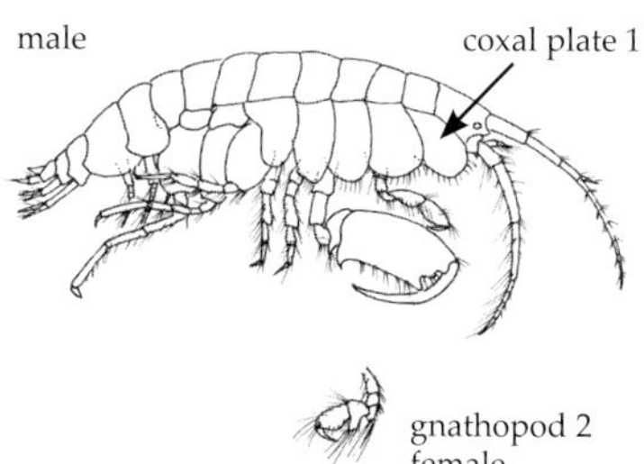

E.39 *Microprotopus maculatus*.

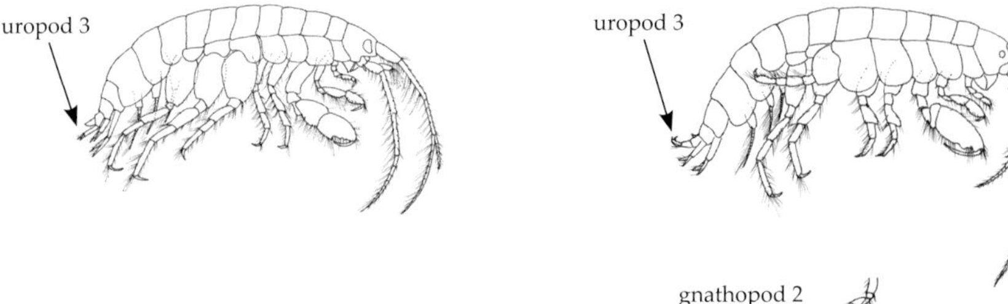

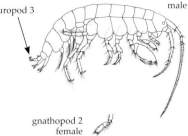

E.40 *Gammaropsis maculata*, female.

E.41 *Sunamphitoe pelagica*.

– Uropod 3 with two almost equal rami. Coxal plates moderately sized, with sparse, stout bristles on edges, but no fine hairs. Antennae up to half body length, with slender, elongate flagellum. Eyes large, oval. Up to 10 mm long; light yellow with dark bands (E.40). Widespread and common. Lower shore, in kelp holdfasts *Gammaropsis maculata*

39. Antennae long, up to half body length. Uropods short; uropod 3 biramous with rami much shorter than peduncle, and two stout hooks at tip of outer ramus. Telson short and broad, with straight edge. Up to 10 mm long; yellow-green, or yellow with red patches (E.41). South and west coasts. Intertidal and shallow sublittoral amongst algae and in kelp holdfasts *Sunamphitoe pelagica*

– Uropod 3 uniramous. Antennae relatively short, less than one-quarter body length; antenna 2 distinctly longer than antenna 1. Pereopods 5–7 increasing in length successively **40**

40. Pereopods 3–7 with large, blunt spine on palm of propodus. Antenna 1 almost as long as antenna 2. Eyes small, oval, pale red. Up to 8 mm long; brownish green (E.42). Widespread and common. Lower shore, in algal turfs and holdfasts *Apohyale prevostii*

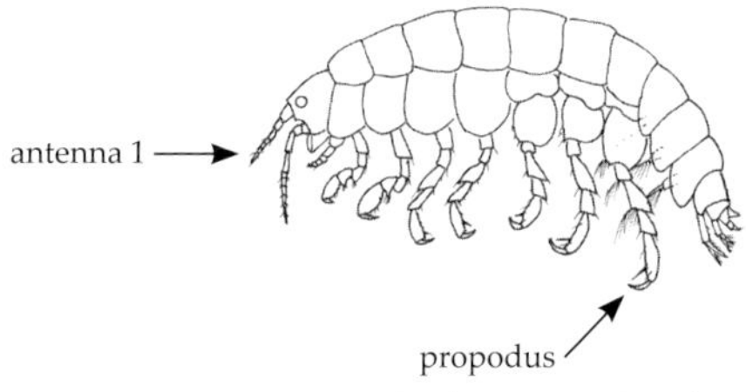

E.42 *Apohyale prevostii*, female.

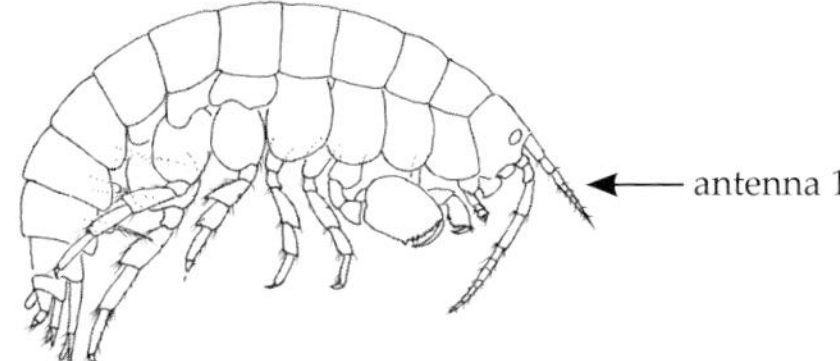

E.43 *Hyale nilssoni*, female.

– Pereopods 3–7 with small spines or bristles on palm of propodus. Antenna 1 a little longer than peduncle of antenna 2. Eyes moderately large, round, black. Up to 8 mm long; brown to green (E.43). Widespread and common. Intertidal, in algal turfs and holdfasts *Hyale nilssoni*

41. Each body segment with a pair of legs (the first two pairs constituting the gnathopods). Slender, oval gill plates at the base of the legs on segments 2, 3 and 4 (E.44) *Phtisica marina*

– Segments 3 and 4 lacking legs, or with minute rudiments. Gill plates on segments 3 and 4 only **42**

42. Tiny, two-jointed legs on body segments 3 and 4. Head and first two segments with erect, forward-directed spines (E.45) *Pseudoprotella phasma*

– No legs on segments 3 and 4 **43**

43. Head bulbous, antenna 2 with tufts of short bristles on lower side (E.46) *Caprella acanthifera*

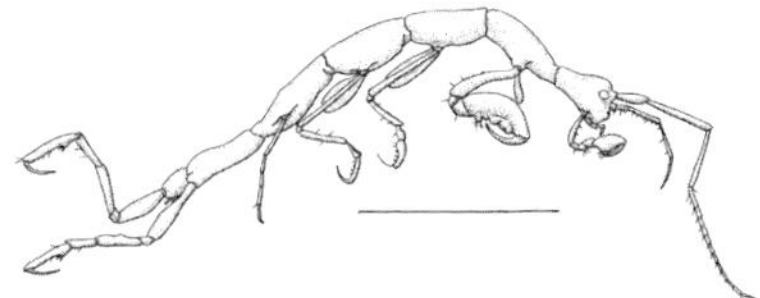

E.44 *Phtisica marina*, male. Scale bar: 10 mm.

E.45 *Pseudoprotella phasma*, female.

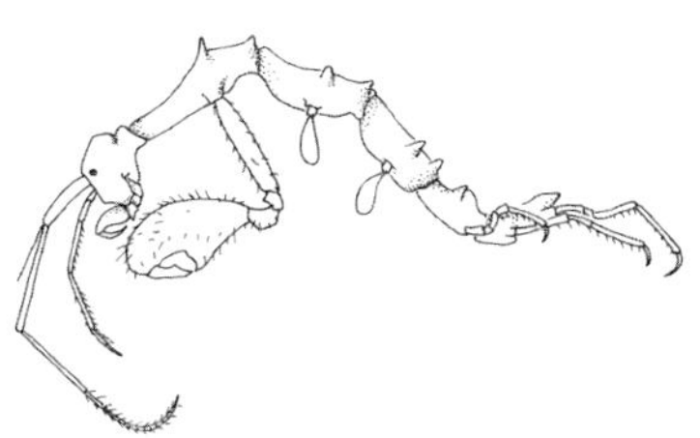

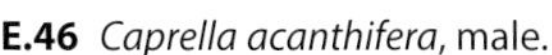

E.46 *Caprella acanthifera*, male.

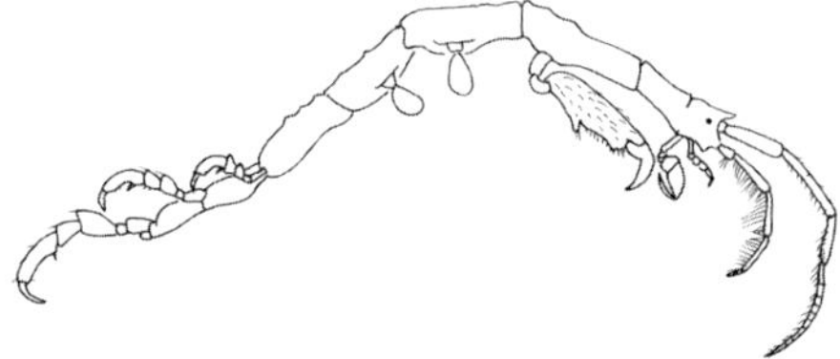

E.47 *Caprella fretensis*, male.

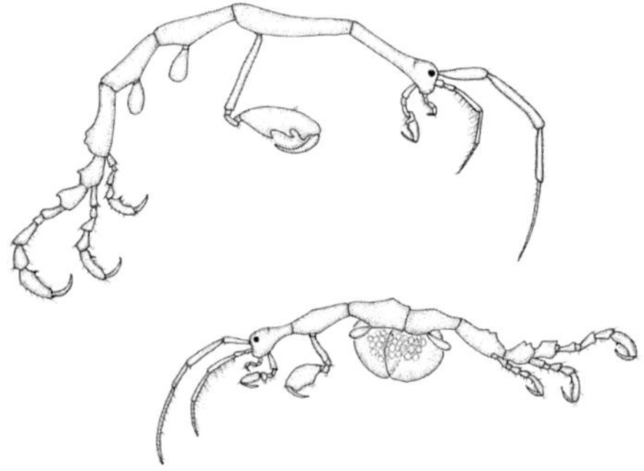

E.48 *Caprella linearis*, male and female (with brood chamber).

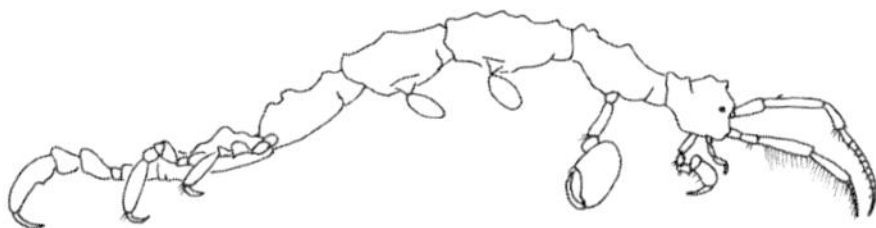

E.49 *Caprella septentrionalis*, male.

– Head more slender (E.47–49). Parallel rows of long bristles on lower side of antenna 2 **44**

44. Head with prominent, forward-directed process in front of eye. Gnathopod 2 arising close to posterior edge of segment 2 (E.47) *Caprella fretensis*

– Head smooth or knobbed, but without a prominent process **45**

45. Head and first body segment, together, about as long as segment 2. Dorsal surface of head and segments 1–4 either smooth, or with a few tubercles in widely spaced pairs (E.48) *Caprella linearis*

– Head and first body segment, together, equivalent to about half length of segment 2. Dorsal surface of head and segments 1–4 typically with irregular, unpaired tubercles (E.49) *Caprella septentrionalis*

Key F Decapods

The Decapoda includes a number of transient members of the rock-pool fauna. In particular, the Common Prawn *Palaemon serratus* moves inshore to breed every spring, and individuals may occur frequently in suitable pools above MTL, while the humpback prawns (*Hippolyte* species) and the Hooded Shrimp *Athanas nitescens* can be found in weedy low-shore pools. *P. elegans* is a smaller species than *P. serratus* and on most coasts is part of the resident rock-pool fauna; it occurs higher upshore than the latter, and shows no seasonal downshore migration, except perhaps on northern coasts. Hermit crabs are also abundant between the tides on rocky shores, and some populations may be permanently resident, moving up- and downshore with the seasons. Brachyuran crabs (those with five pairs of legs visible in dorsal view) also migrate seasonally, with only the smallest individuals remaining low on the shore through the winter, but the anomuran porcelain crabs (four pairs of legs visible) are perhaps resident on the lower shore throughout the year on some coasts. All decapod species occurring around the British Isles may be identified using a number of Linnean Society Synopses: Smaldon *et al.* (1993) for shrimps and prawns, Ingle (1996) for brachyuran crabs, and Ingle & Christiansen (2004) for lobsters, mud shrimps and anomuran crabs. A recent key to the most common species is provided by Ashton *et al.* (2017).

1. Hermit crabs: these need no further description than 'snails with legs'. They will occupy any suitable shell, and in littoral habitats those of winkles (*Littorina* species) and topshells (*Steromphalus* species) are the most abundant available resource. Sublittoral populations have access to a wider range of often larger shells, and the crabs consequently grow larger. *Pagurus bernhardus* (F.1) is

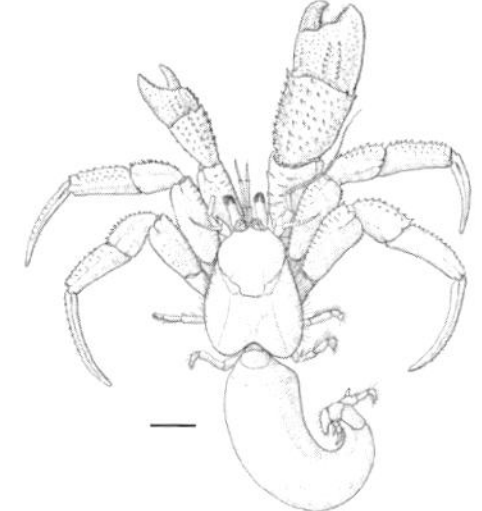

F.1 The hermit crab *Pagurus bernhardus*: **a** with its shell encrusted with the commensal cnidarian *Hydractinia echinata*; **b** entire animal removed from shell (scale bar: 10 mm).

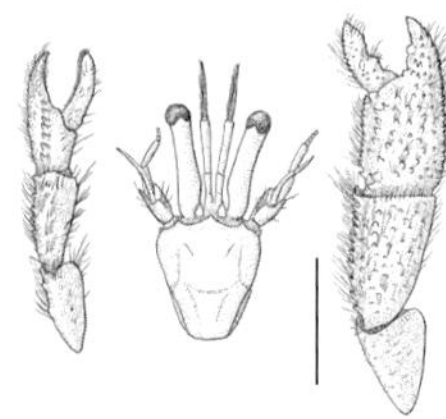

F.2 *Catapaguroides timidus*, cephalothorax and chelipeds. Scale bar: 5 mm.

the most commonly occurring species and is ubiquitous on rocky coastlines of north-west Europe.

Catapaguroides timidus (F.2) is a much smaller species than *Pagurus bernhardus*, its carapace just 5 mm long, and is distinguished from the latter by its slender, more or less equal-sized chelae (claws); it occurs at ELWS on south-west coasts only.

The Mediterranean species *Clibanarius erythropus* also has chelae of almost equal size, both with conspicuous black tips. It formerly occurred as small, isolated populations at sites on the shores of South Devon and Cornwall; they appeared not to have reproduced in the years they were monitored (Southward & Southward 1977, 1988), probably sustained by larval recruitment from populations on the Brittany coast, and were extinct by 1986. However, it was rediscovered at Falmouth in 2016, perhaps suggesting that rising seawater temperatures are stimulating a new colonisation.

– Prawns: laterally flattened decapods (F.3), the anterior cephalothorax bearing five pairs of pereopods, and with a pointed rostrum extending between the eyes; the posterior pleon (abdomen) of six segments, and with five pairs of biramous pleopods and a terminal telson **2**

– Crabs: dorsally flattened, the abdomen folded beneath the carapace and not visible from above; ***or*** with the fifth pereon segment separate from the rest of the cephalothorax and visible from above. First pair of pereopods large, with stout chelae, succeeded by 3–4 pairs of more slender pereopods **8**

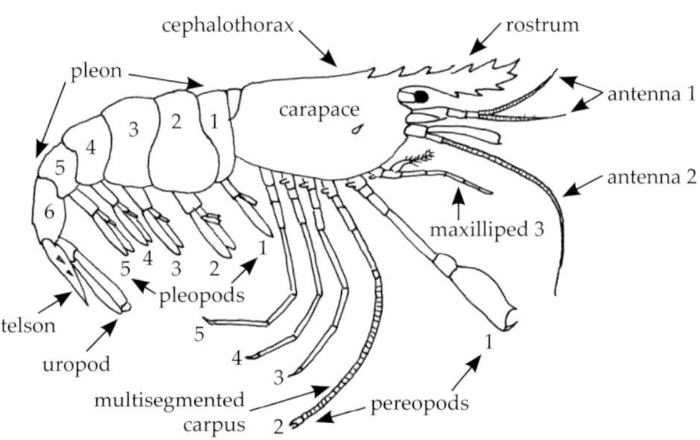

F.3 Morphological features of a prawn.

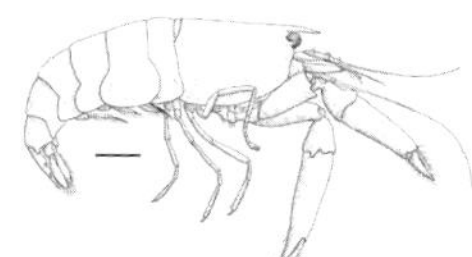

F.4 *Athanas nitescens.* Scale bar: 2mm.

2. Cephalothorax with a very short, single-pointed rostrum, without tooth-like denticulations along its dorsal edge. First pair of pereopods large and stout, with well-developed chelae; pereopods 2–5 much more slender. Up to 20 mm long, blue, green or reddish, with a conspicuous white stripe along the back; resembles a tiny lobster (F.4). Beneath stones, in low-shore pools *Athanas nitescens*

– Cephalothorax with a long, sharply pointed rostrum, typically with sharp teeth along its lower edge, dorsal edge of rostrum and cephalothorax with one to many teeth (lacking in one species). All pereopods similarly long and slender **3**

3. Carpus of pereopod 2 (second walking leg) consists of a single segment. Rostrum and cephalothorax with numerous dorsal teeth **4**

– Carpus of pereopod 2 subdivided into three segments (F.10b). Rostrum and cephalothorax smooth dorsally, or with a single tooth just anterior to the eye socket **7**

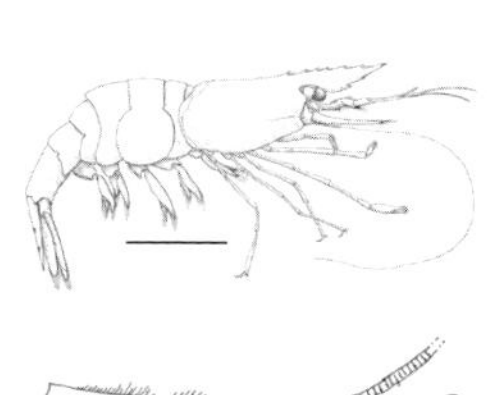

F.5 *Palaemonetes varians*: **a** scale bar: 1 cm; **b** right antennule, to show middle ramus (scale bar: 2 mm).

4. Short middle ramus of the triramous antennule fused for more than half its length with the long outer ramus (F.5b). Rostrum straight, with two teeth ventrally and four to six dorsally, one of which is set posterior to the hind edge of the eye socket. Up to 50 mm long, transparent and mostly colourless (F.5a). Occurs in brackish water pools, often in dense shoals *Palaemonetes varians*

– Short middle ramus of antennule fused for less than half its length with the long outer ramus (F.6b) **5**

5. Mandible with a two-segmented palp (use fine forceps to remove mouthparts from between and below antennae 2 to see the palps) (F.6b). Rostrum straight or with slight upward curve; with seven to nine dorsal teeth, two or three posterior to eye socket, and two to four (commonly three) ventral teeth. Dactylus of pereopod 2 one-third length of propodus. Length around 60 mm; typically with yellow-brown bands (F.6). Intertidal, in rock pools around MTL or higher, perhaps moving downshore in winter on northern shores *Palaemon elegans*

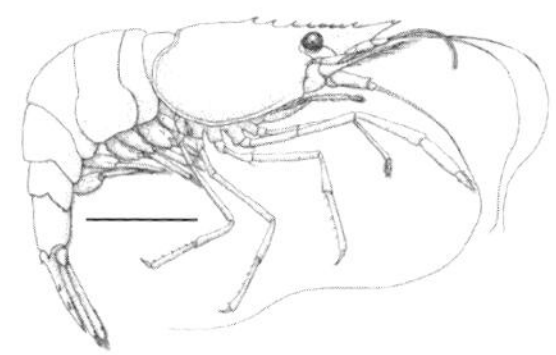

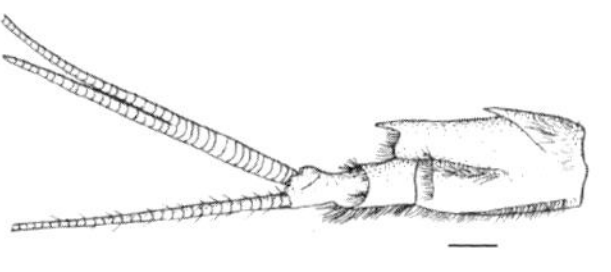

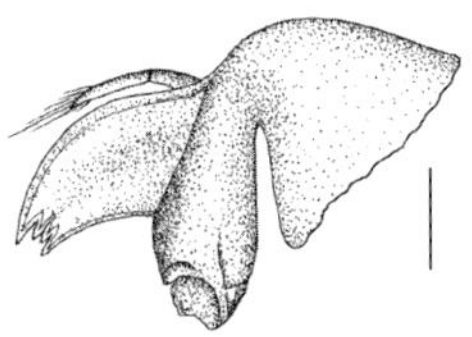

F.6 *Palaemon elegans*: **a** scale bar 1 cm; **b** right antennule, to show middle ramus (scale bar: 2 mm); **c** mandible (scale bar: 1 mm).

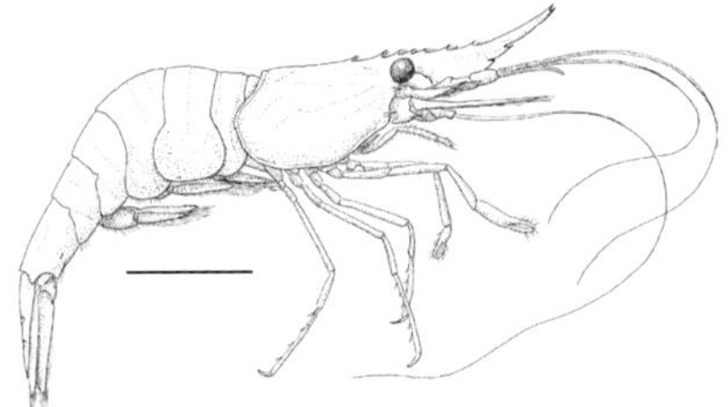

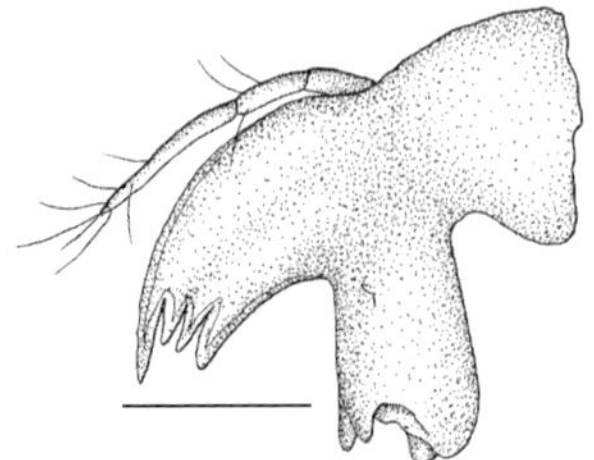

F.7 *Palaemon serratus*: **a** scale bar 1 cm; **b** mandible (scale bar: 1 mm).

– Mandible with three-segmented palp (F.7b). Rostrum straight or strongly upcurved with a sharp bifid point **6**

6. Rostrum strongly upcurved, with sharp, bifid point (i.e. divided in two); with six or seven dorsal teeth, two of which are set posterior to eye socket, but distal third smooth, lacking teeth; four or five teeth ventrally. Pereopod 2 dactylus about half length of propodus. Large, to around 10 cm long, with conspicuous brownish red bands and speckles (F.7) *Palaemon serratus*

– Rostrum straight; commonly with seven to eight (exceptionally to 12) dorsal teeth and three to five ventral teeth; two dorsal teeth set posterior to eye socket, the gap between them greater than between each of the succeeding four teeth. Pereopod 2 dactylus about half length of propodus. Length up to 80 mm; colourless except for faint red speckles (F.8). Estuaries and brackish-water pools, often in large shoals; south and south-east coasts *Palaemon longirostris*

7. Rostrum and carapace smooth dorsally, without teeth; from behind eye socket to its tip, rostrum is longer than the carapace posterior to the socket. Up to 40 mm long; green, less frequently brown or bright red (F.9). Low-shore pools, south and west coasts *Hippolyte inermis*

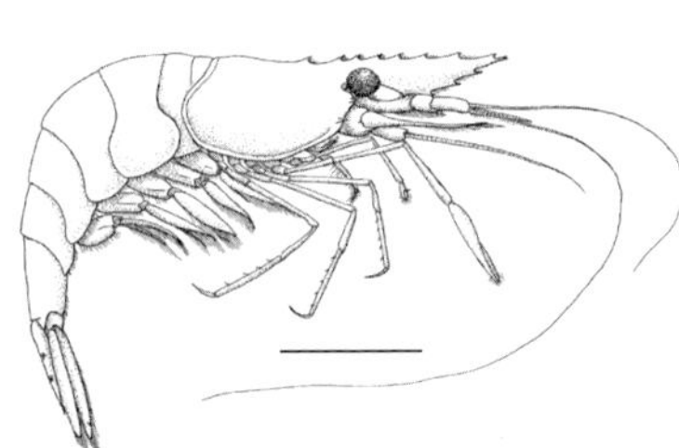

F.8 *Palaemon longirostris*. Scale bar: 1 cm.

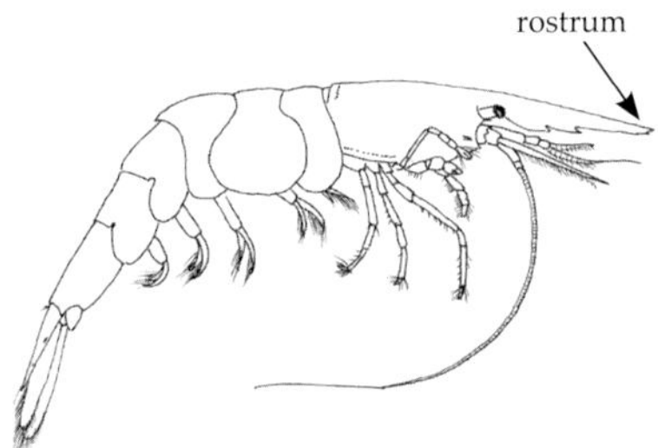

F.9 *Hippolyte inermis*: note smooth dorsal edge of rostrum.

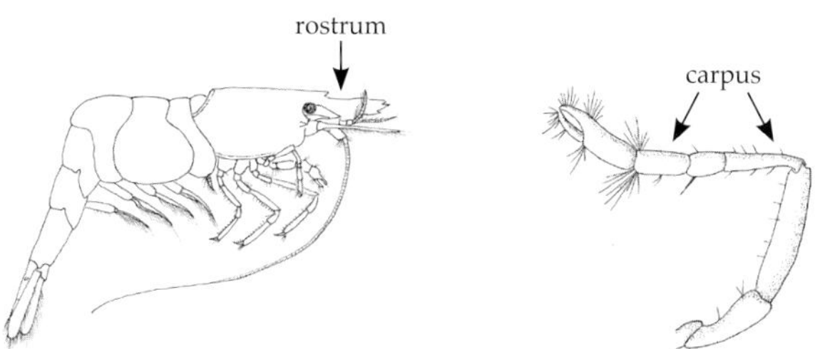

F.10 *Hippolyte varians*: **a** the short, straight rostrum bears a single dorsal tooth anterior to the eye-socket; **b** pereopod 2, showing three-segmented carpus (scale bar: 1 mm).

F.11 A squat lobster *Galathea squamifera*. Scale bar: 10 mm.

– Rostrum with a single dorsal tooth, just anterior to eye socket; its length about equal to length of carapace posterior to socket. Up to 30 mm long. Green, brown or red, with red-brown speckles (F.10). Amongst weed in low-shore pools, all coasts *Hippolyte varians*

8. Four pairs of pereopods visible in dorsal view: a pair of chelae and three pairs of walking legs; fifth pair of pereopods small, and tucked beneath carapace (anomurans) **9**

– All five pairs of pereopods visible from above (brachyurans) **10**

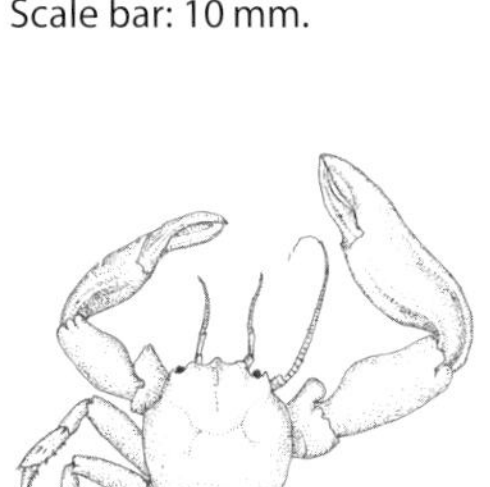

F.12 The porcelain crab *Pisidia longicornis*. Scale bar: 5 mm.

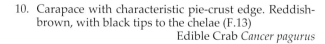

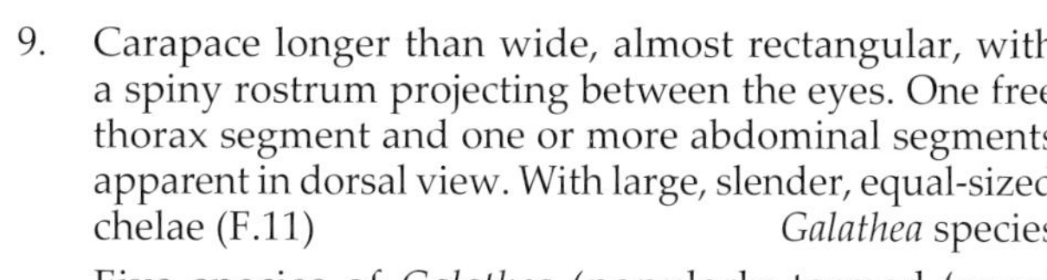

9. Carapace longer than wide, almost rectangular, with a spiny rostrum projecting between the eyes. One free thorax segment and one or more abdominal segments apparent in dorsal view. With large, slender, equal-sized chelae (F.11) *Galathea* species

Five species of *Galathea* (popularly termed 'squat lobsters') occur around the British Isles, two of which, *G. squamifera* (F.11) and *G. nexa*, may be found beneath stones or amongst weed in large, low-shore pools.

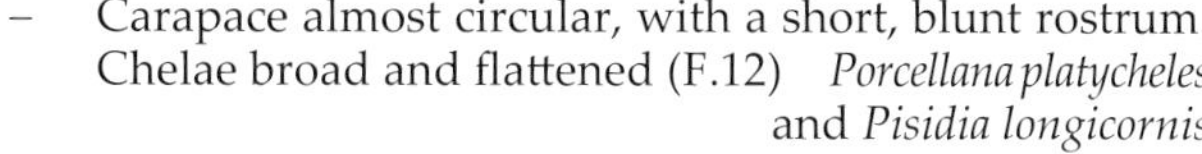

– Carapace almost circular, with a short, blunt rostrum. Chelae broad and flattened (F.12) *Porcellana platycheles* and *Pisidia longicornis*

The two species of porcelain crab are small decapods often common beneath large stones below MTL on sheltered and exposed rocky shores. Both are small animals, with carapaces no more than 15 mm long. *P. longicornis* is quite smooth, while the carapace and appendages of *P. platycheles* are fringed with long setae.

F.13 The Edible Crab *Cancer pagurus* is readily identified by the pie-crust edge of its carapace (photo © Hans Hillewaert/ CC BY-SA 4.0).

10. Carapace with characteristic pie-crust edge. Reddish-brown, with black tips to the chelae (F.13) Edible Crab *Cancer pagurus*

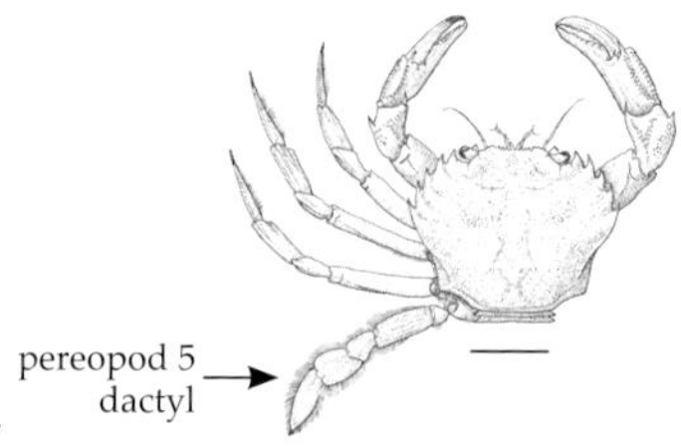

F.14 The Velvet Swimming Crab *Necora puber*. Scale bar: 20 mm.

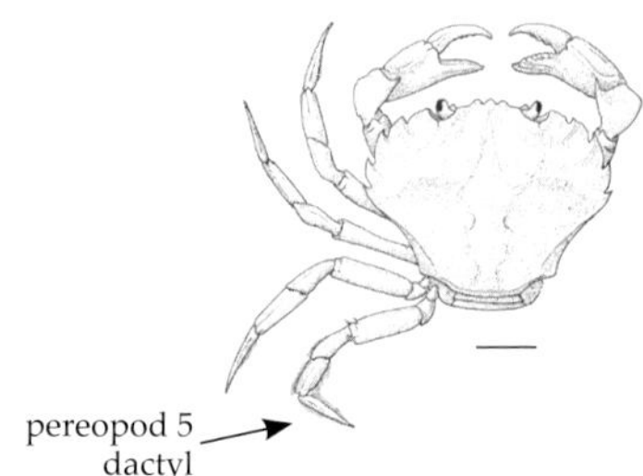

F.15 The Shore Crab *Carcinus maenas*. Scale bar: 10 mm.

Surprisingly large Edible Crabs can be found low on the shore in spring and early summer, almost invariably in deep crevices, often above the water level. **They are best left undisturbed.** Juveniles, with carapace widths below 5 cm, may be common beneath stones, or in crevices, from low water to MTL, and above.

– Carapace with regular series of often sharp denticulations around margin (F.14) **11**

11. Fifth pereopods with propodus and dactyl flattened, in the form of a paddle, fringed with long setae. Bluish, with bright red markings (F.14). Moves rapidly and nips readily (handle with care!)
Velvet Fiddler Crab *Necora puber*

– Fifth pereopods cylindrical distally, pointed, not paddle shaped. Colour variable **12**

12. Carapace smooth, with five sharp marginal teeth on each side, and three short, blunt teeth frontally. Pereopods mostly smooth; few setae distally on pereopod 5 (F.15). Ubiquitous between the tides on all coasts, juveniles ranging upshore to beyond MHWN (see Fig. 2.5)
Shore Crab *Carcinus maenas*

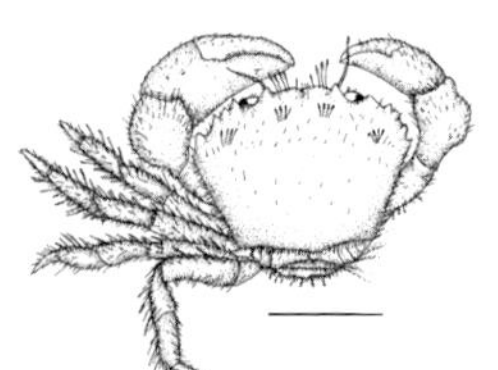

F.16 The Hairy Crab *Pilumnus hirtellus*. Scale bar: 5 mm.

– Carapace with five short teeth on each side, the frontal margin with a tooth close to each eye socket and a lobed edge, covered with short bristles. All pereopods with long and short setae. Reddish brown to purple, up to 15 mm carapace length (F.16). Common below MTL on most rocky shores, beneath stones in larger pools
Hairy Crab *Pilumnus hirtellus*

Key G Shelled gastropods

Limpets, winkles, topshells and dog whelks, representative of three large subclasses of the Class Gastropoda, are the most familiar and conspicuous element of the intertidal molluscan fauna, and are abundantly present on rocky shores across the entire spectrum of wave exposure. Together they will account for the larger part of the molluscan biomass on most shores, and in some instances may account for the larger part of the total of individuals. Yet their combined taxonomic diversity – four species of the limpet genus *Patella*, five winkle species in the genus *Littorina*, three topshells in the genus *Steromphala*, together with *Phorcus lineatus* and the single Common Dog Whelk *Nucella lapillus* – is far outweighed by the, potentially, dozens of species of tiny snails associated with rock-pool algae. None of the large, conspicuous gastropods is especially associated with rock pools; indeed, constant immersion in a pool habitat may lead to an increased risk of predation for most species. Juvenile limpets of all four species can be found on encrusting coralline algae in low-shore pools, from which they move once they have grown to a size at which the risk of desiccation has receded. Upper-shore populations of winkles and topshells may utilise shallow pools as refuges from desiccation during summer emersion, while dog whelks crowd in damp crevices and beneath shaded overhangs. However, as all of these species are likely to be found in the vicinity of pool habitats, they are included in the following key. All species of shelled gastropods found in north-west European shelf environments are described and illustrated in four recent Linnean Society Synopses (Wigham & Graham 2017a, 2017b, 2018; Wigham 2022).

1. Shell conical, without a coiled spire (G.2–G4a) **2**
– Shell with spire of two whorls or more; elongate and spindle shaped; ***or*** short and squat, top or turban shaped, with a depressed spire; ***or*** oval and domed, the spire hidden by the enlarged last (body) whorl, with a slit-like aperture on the lower side (cowries) **7**

2. Shell smooth, or with very faint ridges radiating from the apex of the cone, with blue or pink radiating rays, or brown tortoiseshell patterning **3**
– Shell rugose (wrinkled), with coarse radiating ribs and concentric ridges, the shell margin typically scalloped **5**

G.1 A group of Blue-rayed Limpet *Patella pellucida* on the frond of *Laminaria digitata*.

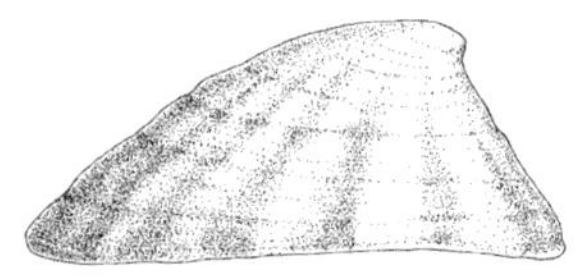

G.2 The shell of the White Tortoiseshell Limpet *Tectura virginea*. Scale bar: 5 mm.

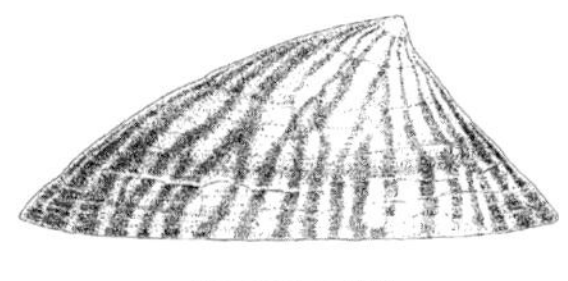

G.3 The shell of the Tortoiseshell Limpet *Testudinalia testudinalis*. Scale bar: 10 mm.

3. Shell translucent, or horn coloured and opaque, with peacock-blue radiating rays. Length up to 10 mm. Blue rays well marked on smaller shells but only showing faintly at apex of largest shells (G.1). On kelp fronds and stipes, or within the holdfast
Blue-rayed Limpet *Patella pellucida*

The Blue-rayed Limpet is principally associated with the seaweed *Laminaria digitata* and is often found as groups of similarly sized individuals, around 6 mm length, at the base of the kelp frond. Larger animals, up to 10 cm long, found within the holdfast have been termed 'variety *laevis*'. The small individuals occur occasionally on other large brown algae, in large, low-shore pools.

– Shell opaque, with pink to brown rays or tessellations. On encrusting coralline algae, never on large brown algae **4**

4. Shell with pink radiating rays. In ventral view, the mantle or pallium (i.e. the fold of tissue overlying the foot and lining the shell) shows a finely papillate (pimpled) edge, with pinkish-red bands (G.2)
White Tortoiseshell Limpet *Tectura virginea*

– Shell with brown tessellations. In ventral view, the mantle edge is green, with short marginal tentacles (G.3)
Tortoiseshell Limpet *Testudinalia testudinalis*

5. Shell regularly conical, with apex more or less central, its rim oval but in ventral view just narrowest at the head end. Foot dull greyish yellow, mantle edge with translucent pallial tentacles (G.4). Common on all rocky shores, from ELWS to MHWN (MHWS on exposed coasts; see Fig. 2.5) Common Limpet *Patella vulgata*

– Shell not a regular cone, apex clearly more anterior, its rim broadest and slightly angular at the posterior end. Foot deep apricot (G.5) or dark grey (G.6), mantle edge tentacles pigmented **6**

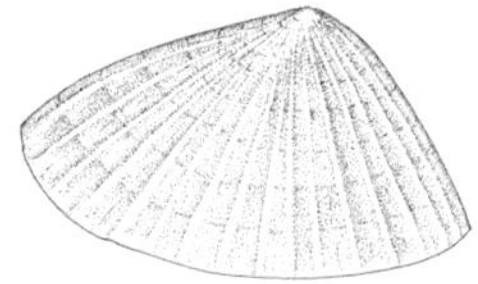

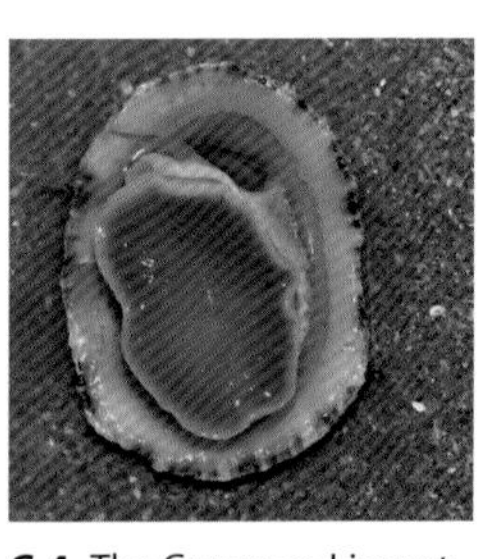

G.4 The Common Limpet *Patella vulgata*: **a** the shell (scale bar: 1 cm); **b** ventral view of live animal to show foot colour. Note that the fold of tissue lining the shell – the mantle – has a translucent fringe.

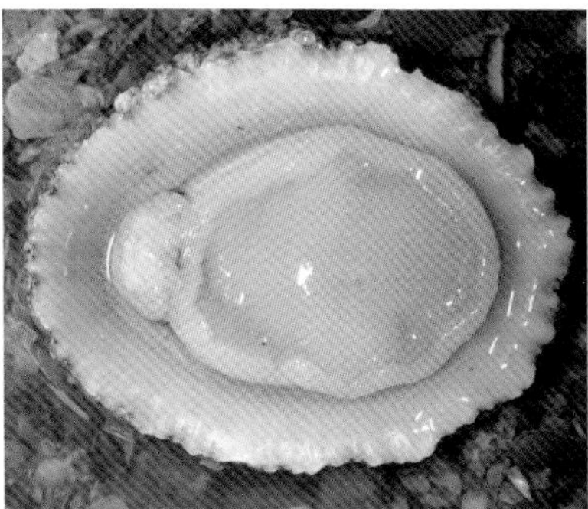

G.5 Ventral view of the China Limpet *Patella ulyssiponensis* to show foot colour. Creamy white tentacles of the mantle fringe are visible above the head (left).

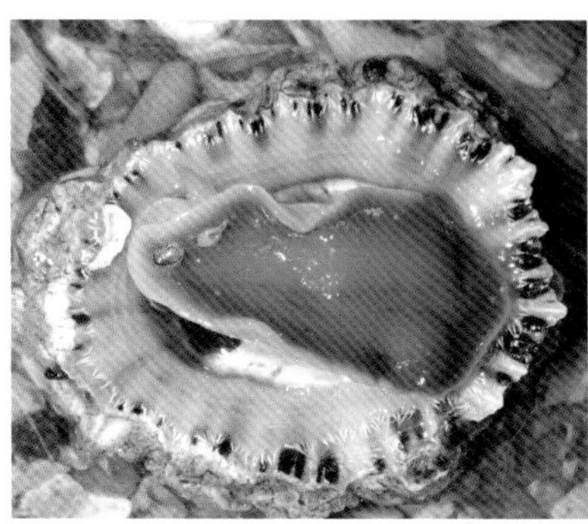

G.6 Ventral view of the Black-footed Limpet *Patella depressa* to show foot colour. Note characteristic chalk-white tentacles fringing the mantle.

6. Foot deep apricot, pallial tentacles rich creamy white (G.5). Lower shore, ELWS to MLWN, intolerant to desiccation and more common on exposed coasts
China Limpet *Patella ulyssiponensis*

– Foot light- to charcoal-grey, pallial tentacles dense chalky white (G.6). Most common from MLWN to MHWN, on wave-exposed shores. South-west coasts of England and Wales Black-footed Limpet *Patella depressa*

G.7 Shells of the Spotted Cowrie *Trivia monacha*, in ventral and dorsal view.

7. Cowries: shell strongly convex, ribbed, with slit-like aperture extending whole length of flat side. Up to 10 mm long. On *Botryllus*, *Botrylloides* and other compound sea squirts encrusting large brown seaweeds. Pinkish white, with three black spots on convex side (G.7)
Spotted Cowrie *Trivia monacha*

The Northern Cowrie *T. arctica* is identical, but lacks spots.

– Not as described **8**

8. Shell with a well-marked groove (siphonal canal) in the lip of the aperture **9**

– Aperture of shell without a siphonal canal **11**

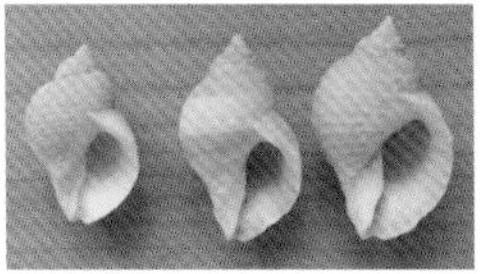

G.8 Shells of the Common Dog Whelk *Nucella lapillus*.

9. Shell with short spire, the broad last whorl (bearing the aperture) comprising about 80% of its total height: up to 40 mm or more. Sculptured with thick spiral ridges (G.8). Common to abundant on all rocky shores
Common Dog Whelk *Nucella lapillus*

– Shell with tall, slender spire, body whorl comprising no more than 40% of total height: up to 7 mm **10**

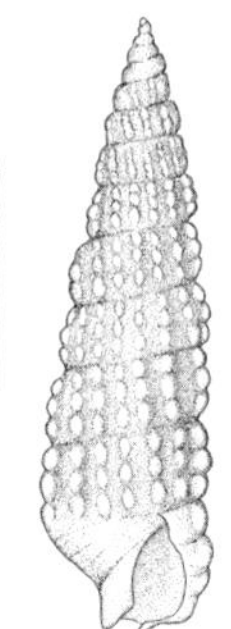

G.9 The shell of *Cerithiopsis tubercularis*. Scale bar: 2.5 mm.

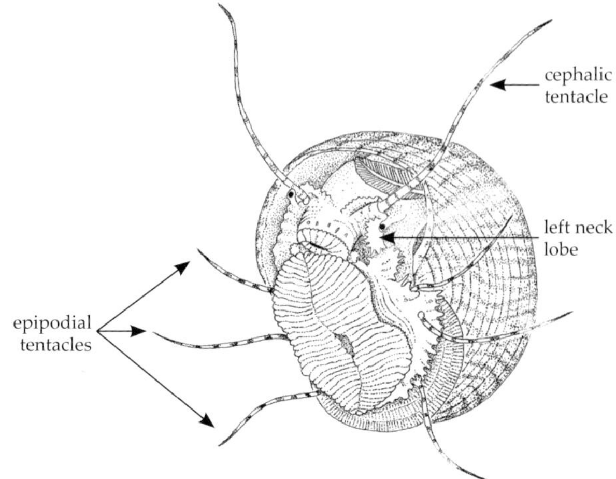

G.11 The Grey Topshell *Steromphala cineraria*: live animal in ventral view. (After Fretter & Graham 1962.)

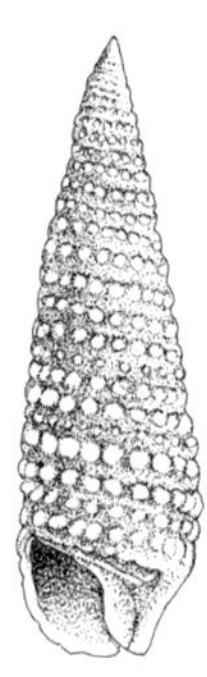

G.10 The shell of *Marshallora adversa*. Scale bar: 2 mm.

10. Aperture on right of long axis, with whorls coiling to the right (dextral); up to 14 whorls, each with three spiral rows of tubercles; to 6.5 mm total height. Chestnut brown (G.9). On sponges, intergrown with *Corallina* and red algal turfs. South-west and west coasts *Cerithiopsis tubercularis*

– Aperture to left of long axis, with whorls coiling to the left (sinistral); up to 15 whorls, each bounded by spiral ridge, with three spiral rows of tubercles on body whorl, decreasing to two towards the apex; to 7 mm total height. Yellowish brown (G.10). In low-shore pools, on sponges, beneath stones or amongst algae. All coasts of Britain and Ireland, except for North Sea and east Channel *Marshallora adversa*

11. Foot with three pairs of epipodial tentacles, and conspicuous neck lobes behind the cephalic tentacles, all visible in crawling animals (G.11). **12**

– Animal lacking both epipodial tentacles and neck lobes **13**

12. Topshells: fat, squat, with reddish or purplish bands across each whorl. Operculum (flap closing shell aperture) brown, transparent. Feeding on algae and algal detritus, often abundant on or beneath bushy fucoids on sheltered rocky shores:

– Coloured bands few, broad. Inner lip of aperture almost straight in basal view. To around 16 mm height, and almost as broad (G.12). Absent from east coasts Purple Topshell *Steromphala umbilicalis*

G.12 The Purple Topshell *Steromphala umbilicalis*: **a** live animals in apical and apertural views; **b** shells in lateral and apertural views.

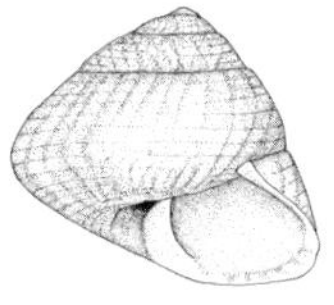

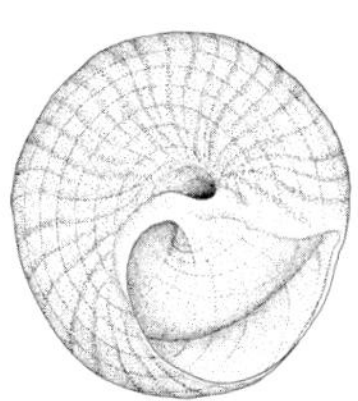

G.13 The Grey Topshell *Steromphala cineraria*: shells in lateral and apertural views (scale bar: 5 mm).

G.14 Shells of the Toothed Topshell *Phorcus lineatus*, in lateral and apertural views. Scale bar: 2 cm.

– Coloured bands numerous, narrow. Inner lip of aperture curved in basal view. To around 16 mm height and breadth (G.13). Common on all rocky coasts of the region
Grey Topshell *Steromphala cineraria*

– Coloured bands developed as irregular mottling. Inner lip of aperture with prominent tooth. To 34 mm height, 30 mm breadth (G.14). South-west coasts only
Toothed Topshell *Phorcus lineatus*

– Shell elongate, glossy, whitish with reddish brown streaks. Operculum thick, white, conspicuous. Up to 7 mm high (G.15). Low-shore pools, particularly associated with small red algae
Pheasant Shell *Tricolia pullus*

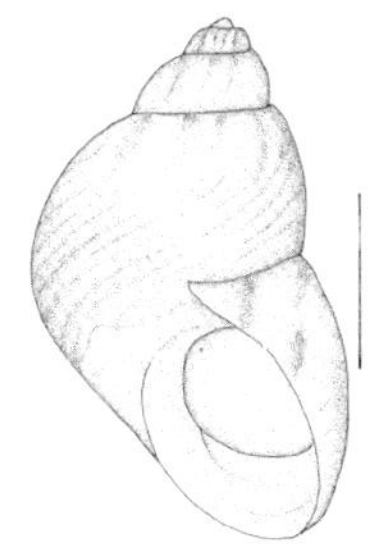

G.15 Shell of the Pheasant Topshell *Tricolia pullus*. Scale bar: 2.5 mm.

13. Shell a minute planospiral (ram's horn shaped), 0.5–1.0 mm diameter. Low-shore pools, amongst filamentous algae **14**

– Not as described, always with a discernible spire **15**

14. Shell with up three rapidly expanding whorls, with clear growth lines but no ribs or ridges (G.16)
Omalogyra atomus

– Shell with three to four gradually expanding, strongly ridged whorls (G.17) *Ammonicera rota*

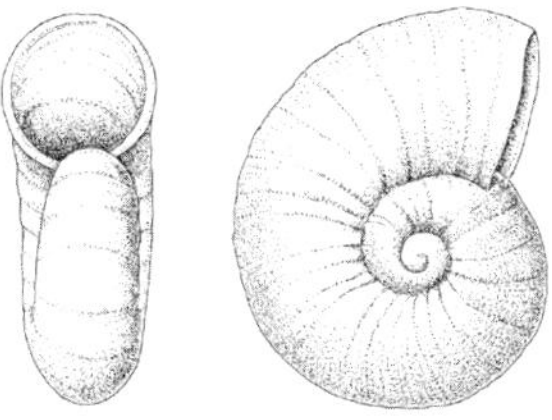

G.16 The shell of *Omalogyra atomus*, in apertural and apical view. Scale bar: 0.5 mm.

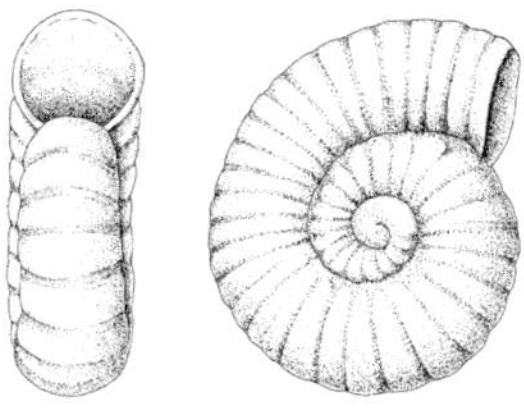

G.17 The shell of *Ammonicera rota*, in apertural and apical view. Scale bar: 0.5 mm.

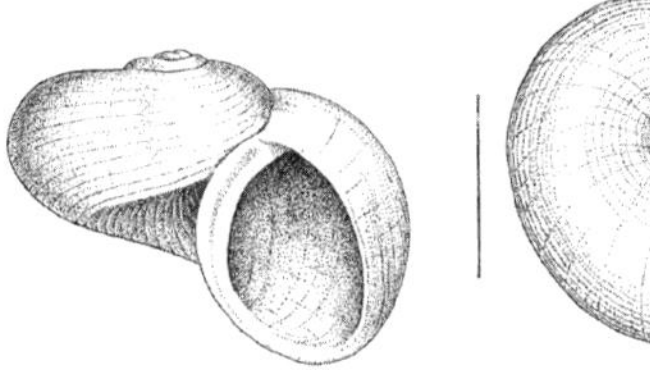
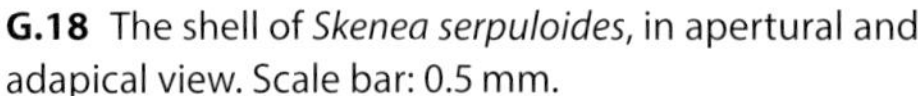

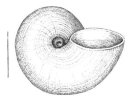

G.18 The shell of *Skenea serpuloides*, in apertural and adapical view. Scale bar: 0.5 mm.

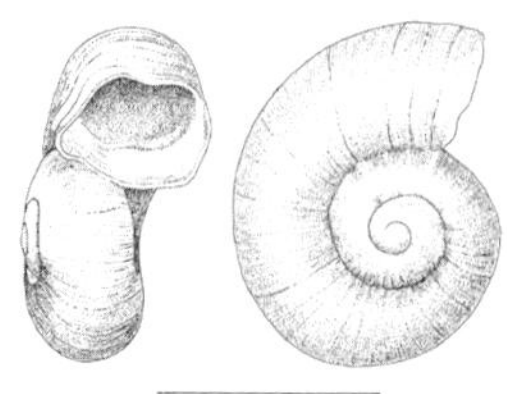

G.19 The shell of *Skeneopsis planorbis*, in apertural and apical view. Scale bar: 1 mm.

15. Shell consists of three to four whorls, the last comprising practically the whole of the shell, defining a deep cavity, the umbilicus, and with a tiny spire contributing about 5% of the total height **16**

– Not as described, spire prominent **17**

16. Shell with growth lines and fine spiral ridges; basal side of last whorl with strong spiral ridges running into the umbilicus. White or colourless, to 2 mm high, 1.5 mm broad (G.18). Most coasts of British Isles except for south-west North Sea and east Channel *Skenea serpuloides*

– Shell almost smooth, with fine growth lines only. Translucent brown, to 0.75 mm high, 1.5 mm broad. Especially in coralline pools (G.19). All coasts except south-west North Sea *Skeneopsis planorbis*

17. Animal with single or paired metapodial tentacle (G.20a, G.27a) **18**

– Animal without metapodial tentacles **26**

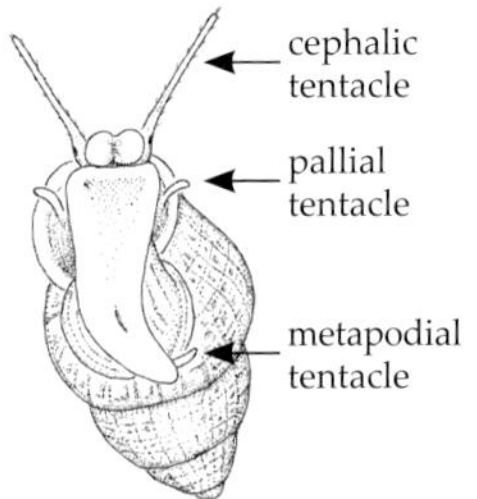

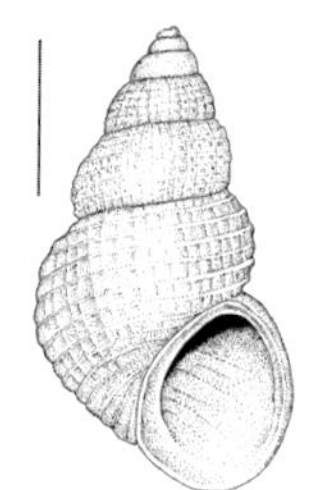

G.20 *Alvania punctura*: **a** in ventral view, showing single metapodial tentacle. (After Fretter & Graham 1962.); **b** shell, scale bar: 1 mm.

18. A single metapodial tentacle, visible at hind end of animal when crawling (G.20a) **19**

– Paired metapodial tentacles, pointing backwards (G.27a) **24**

19. Paired pallial tentacles present, situated just behind head on each side. Shell with spiral and longitudinal ridges, giving a reticulate effect. Up to 3 mm high (G.20). All coasts of the British Isles, except east Channel and south-west North Sea *Alvania punctura*

– Pallial tentacle present on right side only, or absent. Shell smooth, ribbed or grooved, but not reticulate **20**

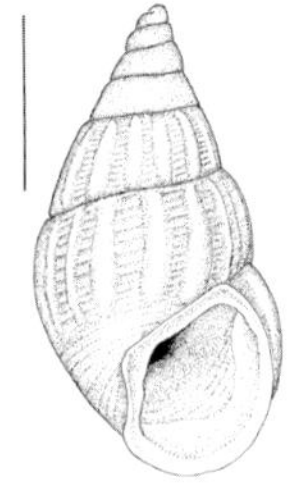

G.21 The shell of *Rissoa lilacina*. Scale bar: 2 mm.

20. No pallial tentacle. Shell with ribs and fine grooves on body whorl only; tip of spire orange; inside edge of outer lip of aperture violet. Up to 4 mm high (G.21). Amongst small red algae on lower shore. South and west coasts *Rissoa lilacina*

– A single pallial tentacle on right side of animal **21**

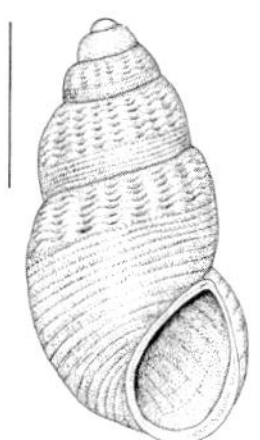

G.22 The shell of *Onoba semicostata*. Scale bar: 1 mm.

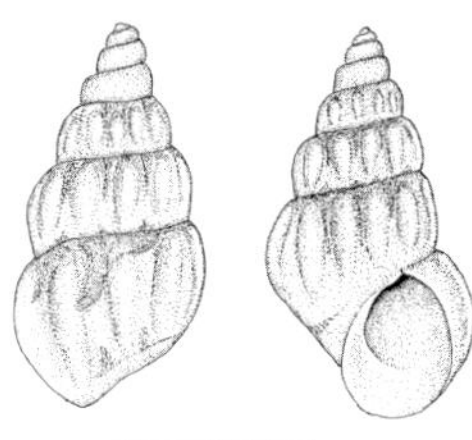

G.23 The shell of *Rissoa parva*. Scale bar: 2 mm.

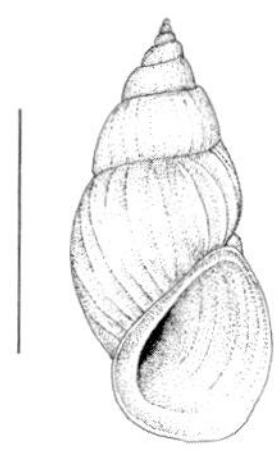

G.24 The shell of *Rissoa membranacea*. Scale bar: 5mm.

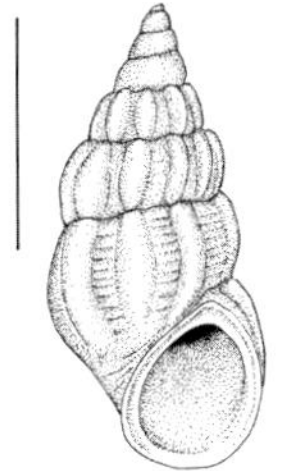

G.25 The shell of *Rissoa guerinii*. Scale bar: 2 mm.

G.26 The shell of *Lacuna pallidula*. Scale bar: 2.5 mm.

21. Shell with fine spiral lines; ribs, if present, developed only on the upper edge of each whorl. Outer lip of aperture without a thickening rib. Up to 3 mm height (G.22). Lower shore, amongst bryozoans and hydroids and algal turfs. Common to abundant on all rocky coasts *Onoba semicostata*

– Not as described **22**

22. Shell smooth or ribbed, with a conspicuous dark comma-shaped mark on upper part of body whorl, close to outer lip. Height up to 4 mm (G.23). Often abundant in filamentous red algal turfs, kelp holdfasts, and tufted bryozoan and hydroid clumps. Common on all rocky coasts *Rissoa parva*

– Not as described **23**

23. Aperture distinctly flared. Body whorl comprising two-thirds height; faint ribs present on body whorl only. Height up to 7 mm (G.24). Low shore, on sheltered coasts; probably mostly south and west *Rissoa membranacea*

– Aperture not flared. Body whorl comprising about half total height; well-developed ribs on all whorls. Height up to 4 mm (G.25). South and west coasts only *Rissoa guerinii*

24. Body whorl expanding broadly, aperture height almost equivalent to total shell height. Up to 8 mm long (G.26). On lower shore fucoids and kelps *Lacuna pallidula*

– Aperture height equivalent to half, or less, shell height **25**

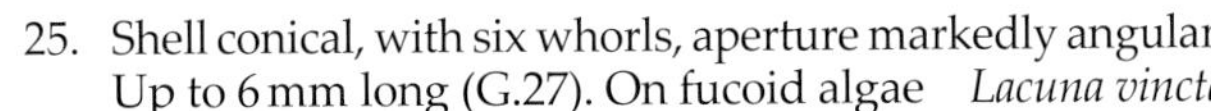

25. Shell conical, with six whorls, aperture markedly angular. Up to 6 mm long (G.27). On fucoid algae *Lacuna vincta*

– Shell globular, with three or four whorls, aperture rounded. Up to 5 mm long (G.28). On *Chondrus crispus* and other small algae. South and west coasts *Lacuna parva*

26. Shell tiny, to 1.5 mm long, transparent; conical, with 4.5 whorls (G.29a). Cephalic tentacles cylindrical. On filamentous red and small green algae, in coralline rock pools from MTL to ELWS; south-west and west coasts of Britain, west coasts of Ireland *Rissoella diaphana*

Two other species occur in similar habitats: *R. opalina* (G.29b), up to 2 × 1.4 mm, has a globular shell with the body whorl comprising more than three-quarters its total length and bifid cephalic tentacles; *R. globularis* is minute, only up to 1 mm length, and quite globular, with cylindrical cephalic tentacles.

– Not as described **27**

27. Shell minute (1 mm long), pale coloured, with spiral brown bands (G.30). On red and green algae in rock pools. South and west coasts *Eatonina fulgida*

– Not as described **28**

28. Shell elongate, dark red, or with broad, dark red bands. Up to 3 mm long. Operculum deep crimson, with

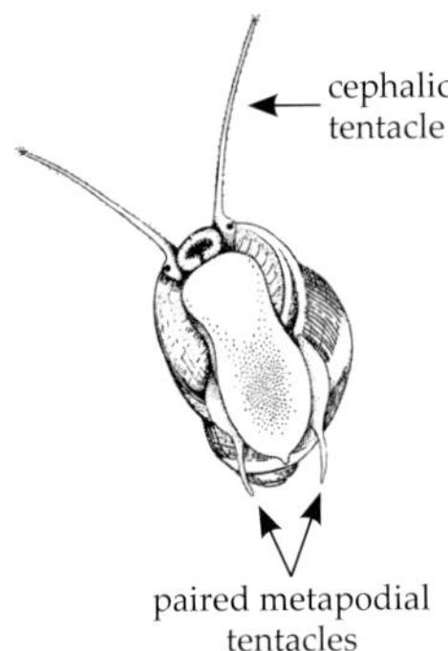

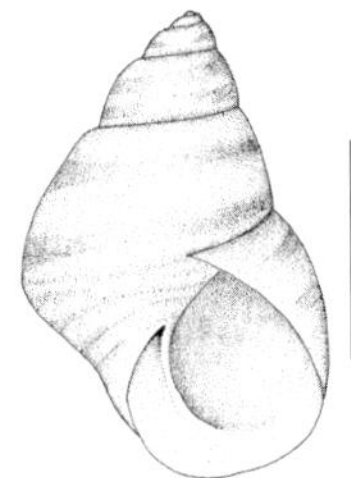

G.27 *Lacuna vincta*: **a** in ventral view, showing paired metapodial tentacles. (After Fretter & Graham 1962.); **b** shell (scale bar: 2.55 mm).

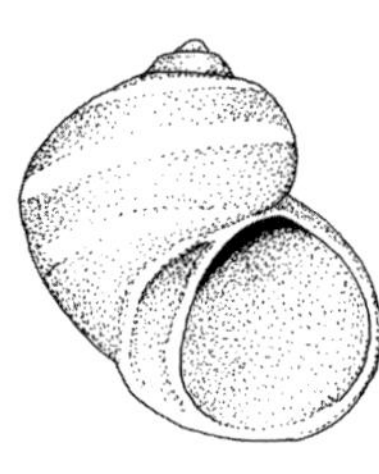

G.28 The shell of *Lacuna parva*. Scale bar: 2 mm.

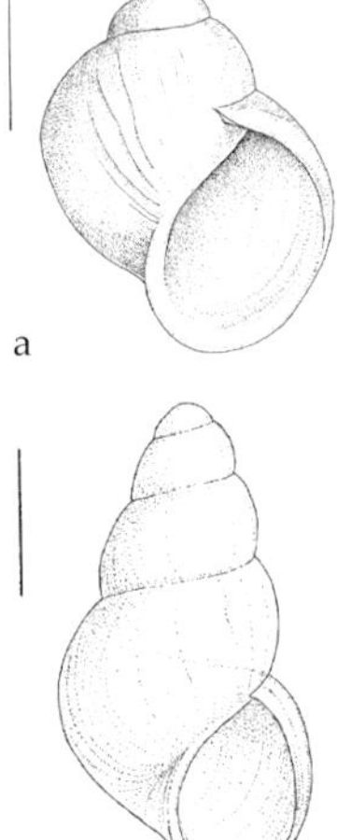

G.29 The shells of **a** *Rissoella diaphana* and **b** *R. opalina*. Scale bars: 0.5 mm.

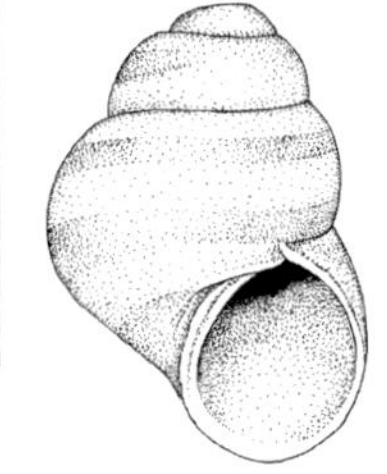

G.30 The shell of *Eatonina fulgida*. Scale bar: 1 mm.

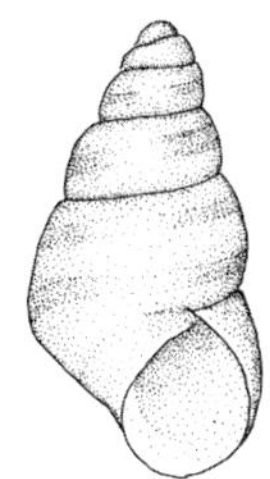

G.31 The shell of *Barleeia unifasciata*. Scale bar: 2 mm.

concentric lines (G.31). Lower shore, in red algal turfs. South and west coasts *Barleeia unifasciata*

– Shell squat, exceeding 3 mm long. Operculum horny, with spiral lines **29**

29. Shell with pointed spire, ornamented with spiral lines and grooves. Aperture flared. Colour variable *Littorina* species

The Common Periwinkle *Littorina littorea* (G.32) is common to abundant on all but the most wave-exposed rocky shores. The Rough Periwinkle *L. saxatilis* (G.33) is equally abundant, and occupies a wider range of habitat, from mudflats to rocky shores, and an extended tidal range, occurring in high supralittoral pools. Both species may occur sympatrically with *L. compressa*, the shell of which bears flat spiral ridges, often with black pigment between.

– Shell globose, spire depressed; aperture as wide as shell height **30**

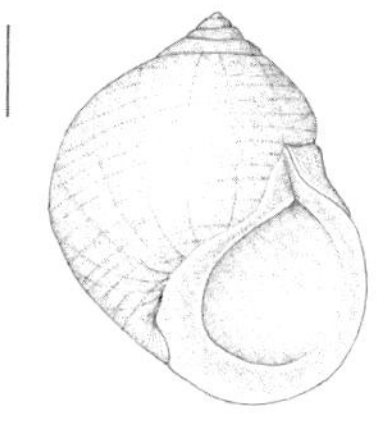

G.32 The shell of the Common Periwinkle *Littorina littorea*. Scale bar: 5 mm.

30. Aperture relatively broad, outer lip arising well below level of apex. Shell with obliquely oval outline in apertural view (G.34). Middle to upper shore, on sheltered to moderately exposed coasts, typically associated with *Fucus vesiculosus* and *Ascophyllum nodosum*. Common on all coasts in the region *Littorina obtusata*

– Aperture relatively narrow, outer lip arising close to apex. Shell with obliquely drop-shaped outline in apertural view (G.35). Middle to lower shore, on sheltered to moderately exposed coasts, typically associated with *Fucus serratus* *Littorina fabalis*

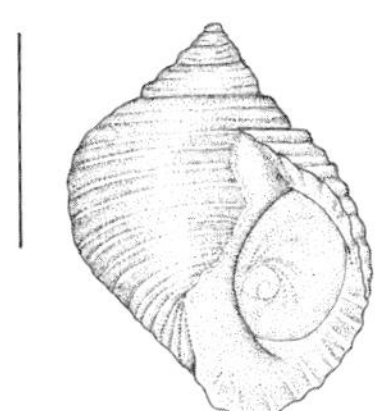

G.33 The shell of the Rough Periwinkle *Littorina saxatilis*. Scale bar: 5 mm.

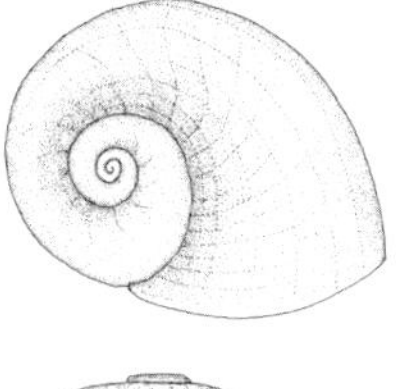

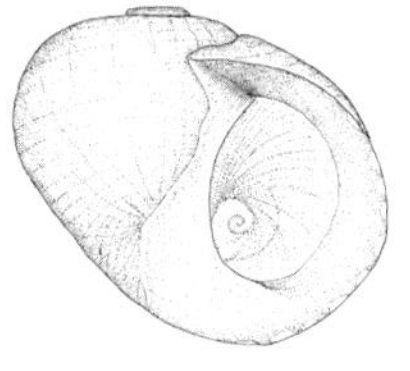

G.34 The shell of *Littorina obtusata*. Scale bar: 5 mm.

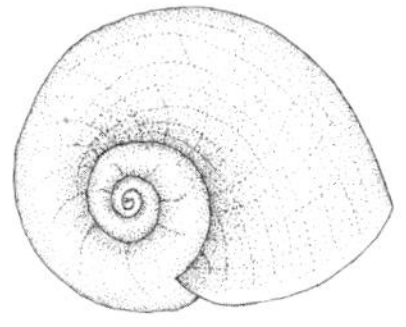

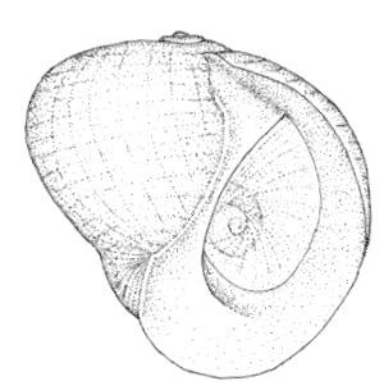

G.35 The shell of *Littorina fabalis*. Scale bar: 5 mm.

Key H Chitons

The chitons constitute one of the smaller molluscan classes, the Polyplacophora. They are all epilithic, herbivorous grazers; like the limpets (*Patella* species), they employ an extensive muscular foot to clamp onto their substratum. Chitons are oval, the head, foot and viscera being enclosed by a series of eight, interlocking, arched plates, bordered by a fold of flexible tissue (the girdle). In contradistinction to the limpets, in which the rim of the shell fits the contours of the individual's home site, this girdle seals the underside of the animal and its mantle cavity against the exterior. North-east temperate Atlantic chiton species tend to be small, morphologically conservative and of low taxonomic diversity. They are rather well camouflaged on their rocky substrata and easy to overlook. None are especially associated with rock-pool habitats, although *Boreochiton rubra* and *Tonicella marmorea* are most often found attached to sheet-encrusting coralline algae. The chiton species most commonly recorded in the region are described and illustrated in Hayward & Ryland (2017); a more comprehensive account of British species was provided by Jones and Baxter (1987).

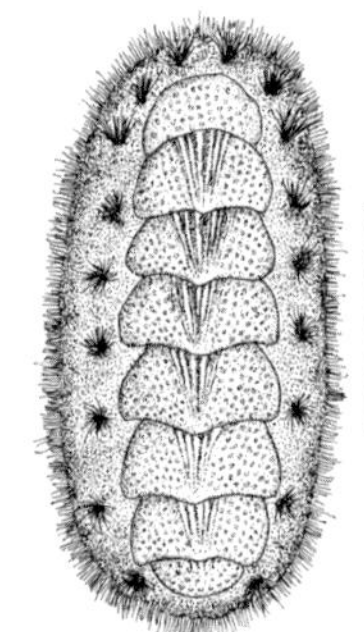

H.1 *Acanthochitona crinita*. Scale bar: 5 mm.

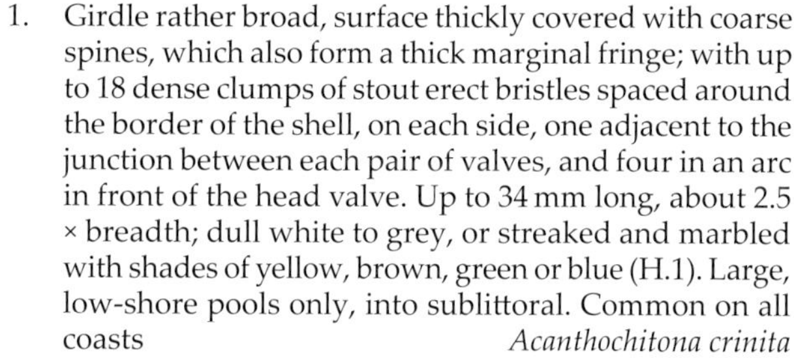

1. Girdle rather broad, surface thickly covered with coarse spines, which also form a thick marginal fringe; with up to 18 dense clumps of stout erect bristles spaced around the border of the shell, on each side, one adjacent to the junction between each pair of valves, and four in an arc in front of the head valve. Up to 34 mm long, about 2.5 × breadth; dull white to grey, or streaked and marbled with shades of yellow, brown, green or blue (H.1). Large, low-shore pools only, into sublittoral. Common on all coasts *Acanthochitona crinita*

   Acanthochitona fascicularis is a larger species, up to 60 mm long, and is particularly distinguished by the sculpture of it shell valves, which is much finer than the coarse tuberculation seen in *A. crinita*. It is restricted to south and west coasts of England and Wales, and to western Ireland.

– Girdle smooth, granular or scaly, with or without a marginal fringe of fine spines, but without clumps of erect bristles **2**

2. Girdle narrow, covered with flat scales and with a fringe of fine spines **3**

– Girdle broad or narrow, smooth, or covered with granules, sometimes with recumbent spines, and with a marginal fringe of fine spines **4**

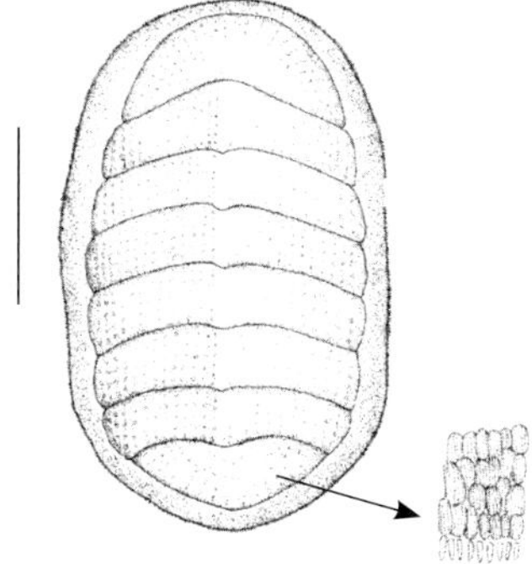

H.2 *Leptochiton asellus*, with detail of girdle edge. Scale bar: 5 mm.

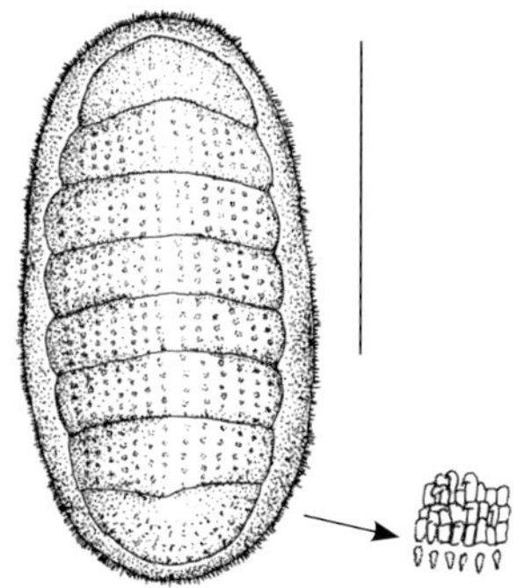

H.3 *Leptochiton cancellatus*, with detail of girdle edge. Scale bar: 5 mm.

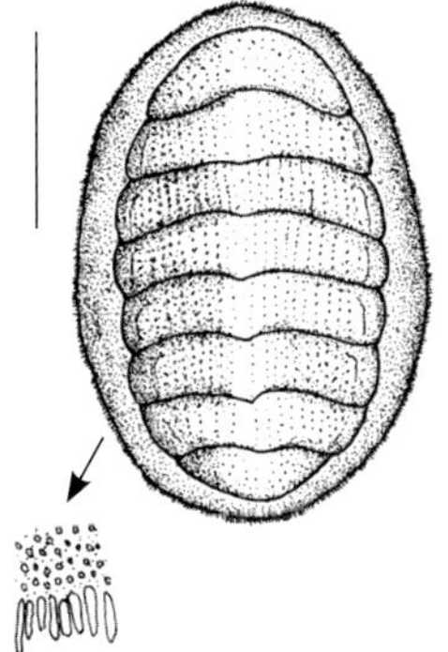

H.4 *Lepidochitona cinerea*, with detail of girdle edge. Scale bar: 5 mm.

3. Each shell valve with a pronounced keel, extending posteriorly as a prominent beak. Girdle covered with flat, striated scales. Up to 18 mm long, about 1.5 × breadth, dull white to yellow, often with dark encrustations (H.2). Common on all coasts *Leptochiton asellus*

– Shell valves smoothly arched without keels or prominent beaks. Small, up to 9 mm long, about 2 × breadth, pale yellow-white to fawn (H.3). Most frequent on south and west coasts *Leptochiton cancellatus*

4. Shell valves finely granular. Girdle narrow, covered with coarse, variably sized granules, and with a fringe of short, blunt spines. Up to 24 mm long; red, brown, yellow to green, in streaks and patches (H.4). May occur in pools from MTL and below. Common on all coasts *Lepidochitona cinerea*

– Shell valves smooth and glossy, marbled with shades of brown and white. Girdle with similar coloration and patterning, densely covered with small spherical granules and fringed with small, flat spines. Up to 20 mm long (H.5). In low-shore pools *Boreochiton rubra*

Tonicella marmorea is similar but larger, up to 40 mm long. The girdle granulation is sparse, and so fine that the girdle appears quite smooth. Northern shores, from west Scotland to Yorkshire.

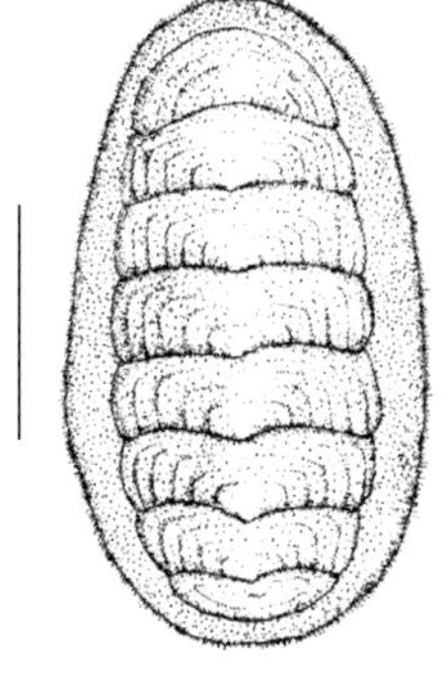

H.5 *Boreochiton rubra*. Scale bar: 5 mm.

Key I Bivalves

In the north-east Atlantic region, species of the molluscan class Bivalvia are most frequently encountered in unconsolidated, sedimentary environments, from muddy sands to coarse gravels, as part of the infauna, or as surface-living epifauna. Some epifaunal species, such as the Horse Mussel *Modiolus modiolus*, create dense and extensive beds, which show long-term persistence, and may be considered as significant ecological engineers. A minority of species, such as the scallops, has adopted a motile lifestyle while others, such as the saddle oysters, have adapted to a completely sessile existence, cemented to hard substrata. The few species commonly found in rocky intertidal habitats tend to be crevice dwellers or rock borers, or sedentary in habit, attaching securely to the substratum by elastic byssal threads, secreted by a gland in the animal's foot. *British Bivalve Seashells* (Tebble 1976) remains the most useful practical aid for the identification of species found in British shelf waters. A comprehensive review of the British bivalve fauna, including detailed descriptions, notes on their biology and ecology, and high-quality digital images can be found online in Oliver *et al.* (2010).

pallial line
a line on the inner surface of each shell valve running between the two adductor muscle scars and marking the attachment of the mantle edge to the shell; a posterior indentation, the pallial sinus, indicates the position of the retracted siphon

In bivalves the plane of symmetry passes between the two shell valves, and it is necessary to distinguish left from right. Thus, the hinge line is dorsal, the opposing shell margin ventral; umbones are anterior to the vertical midline of the shell; shells may be more or less inequilateral, when the anterior portion and the anterior muscles scars are the smaller part; the pallial sinus, when present, is always posterior (I.7–I.11).

umbo (pl. umbones)
the earliest part of the shell, corresponding to the first-formed larval shell

1. Shell with umbones at or near one end; rounded-triangular, oval or bean shaped, and inner edges without hinge teeth (I.1) **2**
– Umbones not at or near end, hinge teeth present (I.7–I11) **7**

2. Shell smooth, with or without concentric or radiating lines, but without raised ribs **3**
– Shell with two series of radiating ribs, separated by a smooth area, extending from the umbones to the lower margin **5**

3. Umbones at the extreme (anterior) end of the shell. Blue-black coloration. Commonly 50–100 mm long, but no more than 30 mm in some populations and up to 150 mm in others; size and shape determined by

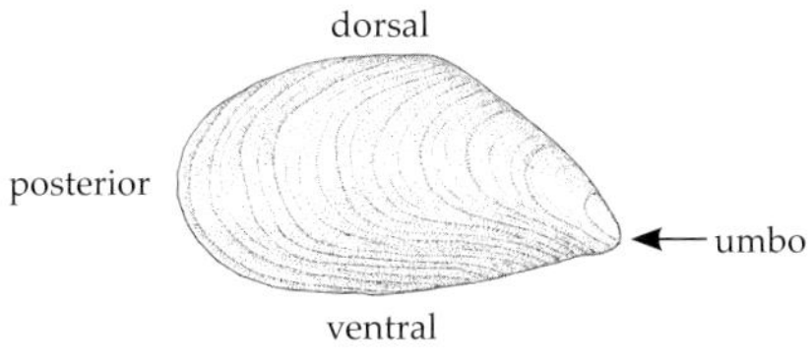

I.1 The Common or Blue Mussel *Mytilus edulis*: right-hand view.

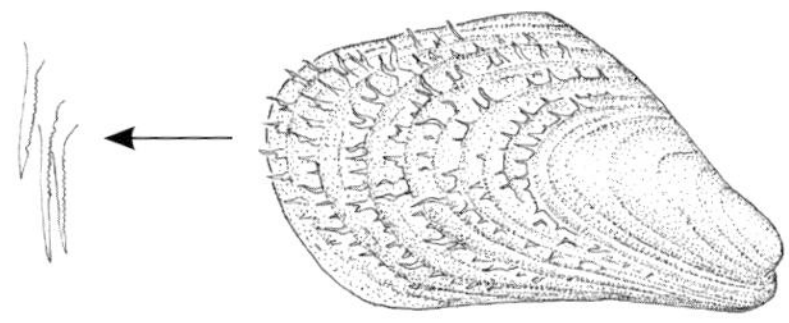

I.2 *Modiolus barbatus*: right-hand view, detail shows toothed periostracal spines.

environmental factors, especially wave exposure (I.1). ELWS to MHWN, higher on exposed shores. Post-larvae typically settle on fine filamentous algae, migrating via byssal drifting to final habitat. Favours open rock, holdfasts of large, lower shore algae, high-shore coralline pools Common or Blue Mussel *Mytilus edulis*

byssus
a bundle of golden-yellow fibres, tough and elastic, secreted by a gland in the animal's foot; produced only by the post-larva and acting as a drift anchor, or serving to attach the adult to its substratum

– Umbones not quite at end of the shell, its shell margin extending beyond the beaks **4**

4. Shell light yellowish red, covered by darker yellow brown outer skin (periostracum) bearing fringes of broad, saw-toothed spines. To 60 mm long (I.2). Lower shore, among kelp holdfasts. South coasts, to Yorkshire on the east, to Clyde Sea on the west *Modiolus barbatus*

– Shell white, bluish or purple, periostracum light yellow brown in juveniles, deep mahogany in larger individuals. Some smooth (not saw-toothed) periostracal spines in small specimens. Up to 100 mm long, much larger in offshore populations (I.3). Occasional around ELWS, amongst kelp holdfasts, abundant on mixed coarse grounds offshore Horse Mussel *Modiolus modiolus*

5. Umbones very prominent, shell rather plump. Dorsal margin downcurved posteriorly; up to 18 anterior ribs, 20–35 posteriorly. Horn coloured or light green, often with brown or red-brown markings. Up to 20 mm long (I.4). Attached by byssus to holdfasts of *Laminaria* and other large lower shore algae *Musculus marmoratus*

– Umbones distinct, but not prominently so, shell slender. Dorsal margin upcurved, the posterior end as deep as the midline **6**

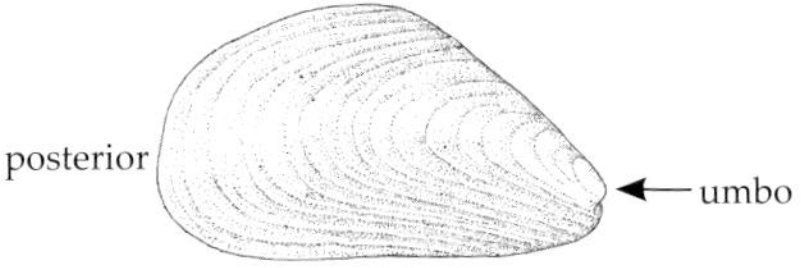

I.3 The Horse Mussel *Modiolus modiolus*: right-hand view.

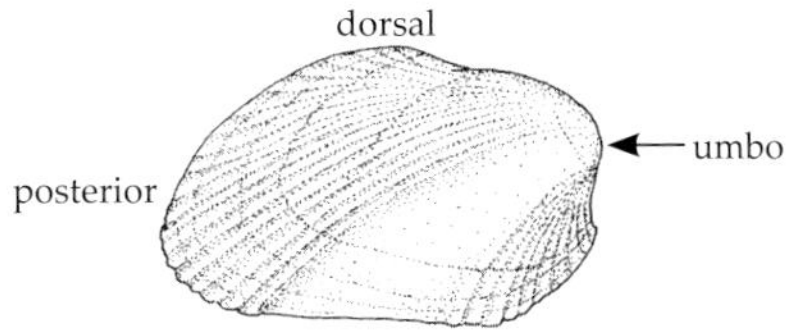

I.4 *Musculus marmoratus*: right-hand view.

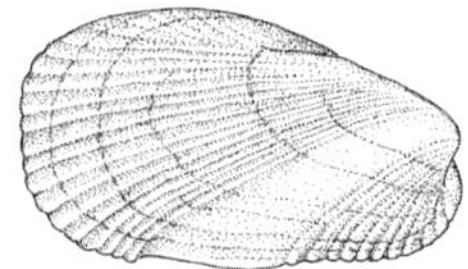

I.5 *Musculus costulatus*: right-hand view.

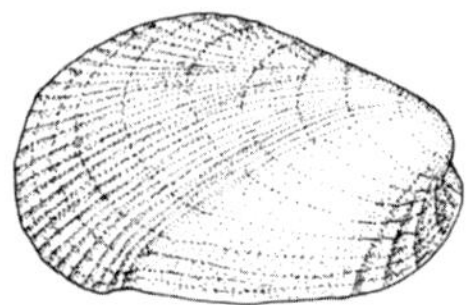

I.6 *Musculus discors*: right-hand view.

6. Shell elongate oval, 10 broad anterior ribs, 20–30 slender ribs posteriorly. Light or dark greenish brown with streaks and blotches of green and reddish brown, often in chevrons along hinge line. Up to 13 mm long (I.5). Attached by byssus to holdfasts of *Laminaria* and other large lower shore algae *Musculus costulatus*

– Shell oval or rhomboidal, eight to 12 broad anterior ribs, 30–45 slender ribs posteriorly. Pale green or brown, often the same colour all over, or with a few darker patches. Up to 13 mm long (I.6). Attached by byssus amongst holdfasts of large, lower-shore algae, in bryozoan turfs and amongst filamentous algae *Musculus discors*

7. Small, plump, almost circular, yellowish white, up to 10 mm long. Right valve with two separate hinge teeth; left valve with one pair plus one single. Pallial line without a sinus (I.7). A crevice-dwelling species, sometimes occurring in algal holdfasts *Kellia suborbicularis*

– Not as described **8**

8. Flat, oval, yellowish brown, tinged with pink or red, darkest at the umbones, up to 3 mm long. Right valve with single anterior and posterior lateral teeth; left valve with one small hinge tooth below the umbo and single anterior and posterior lateral teeth. Pallial line without a sinus (I.8). Intertidal, attached by byssus, in crevices and, often in large numbers, in *Lichina pygmaea* and upper shore fucoid algae *Lasaea adansoni*

– Not as described **9**

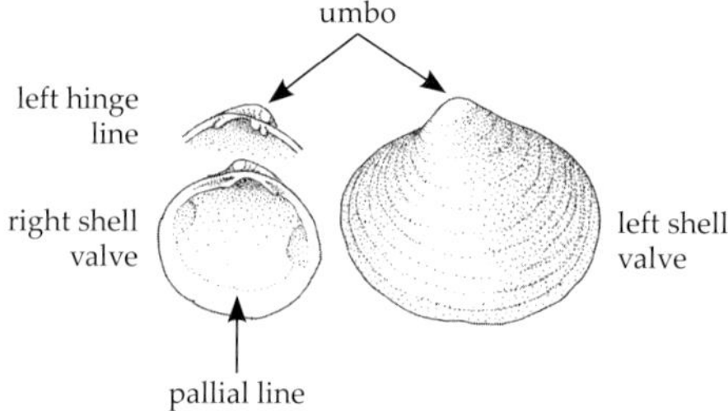

I.7 *Kellia suborbicularis*: left-hand view of exterior, interior view of right valve, hinge line of left valve.

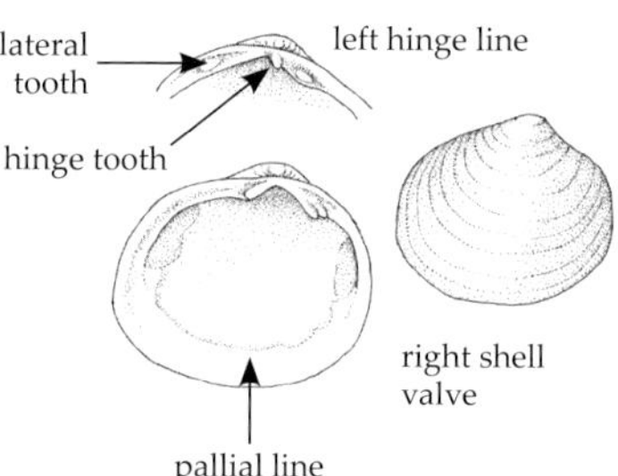

I.8 *Lasaea adansoni*: right-hand view of exterior, interior view of right valve, hinge line of left valve.

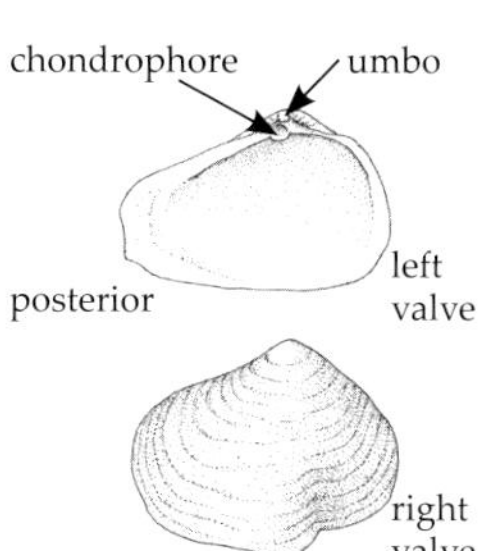

I.9 *Sphenia binghami*: interior view of left valve, exterior view of right.

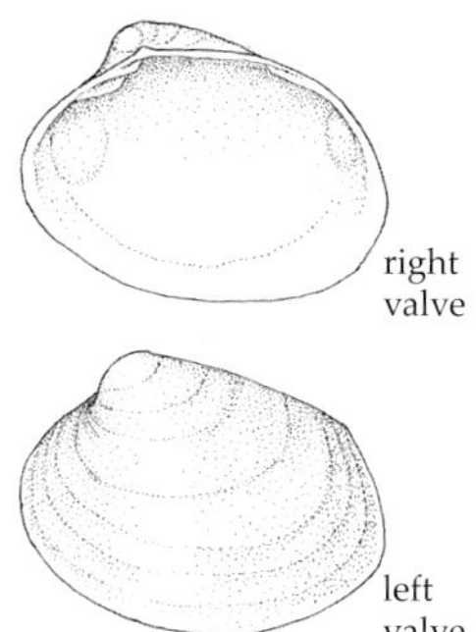

I.10 *Turtonia minuta*: interior view of right valve, exterior view of left.

I.11 *Irus irus*: interior view of right valve, exterior view of left.

9. Hinge ligament internal and external, the internal portion set in a conspicuous pit (the chondrophore). No hinge teeth. Shell irregularly oval, up to 20 mm long (I.9). Attached by byssus in algal holdfasts on lower shore *Sphenia binghami*

– Hinge ligament external only, no chondrophore. Each shell valve with three hinge teeth **10**

10. Shell oval, smooth, plump, not exceeding 3 mm long. Dull white to light brown (I.10). Intertidal only, attached by byssus, in crevices, often abundant in algal holdfasts *Turtonia minuta*

– Shell elongate, oblong, with conspicuous concentric ridges, up to 25 mm long (I.11). Yellow white to light brown. Lower shore, in crevices and kelp holdfasts *Irus irus*

Key J Polychaetes

Numerous species of polychaete worm are likely to be found in mid- to low-shore rock pools, although their diversity is not always readily apparent, and there seem to be no studies focused especially on rock-pool polychaete communities. The spirorbid tubeworms are conspicuous, encrusting coralline algae and some of the larger brown algae, and their aggregated populations have attracted considerable research effort (e.g. Knight-Jones & Knight-Jones 1977), but other species are less conveniently distributed. Some occupy silty habitats, in crevices or beneath stones, but perhaps a majority of polychaete species in the rocky intertidal are associated with algae. Algal holdfasts provide habitat for both tube-building and free-living worms, and food in the form of detritus, micro-organisms and associated meiofaunal prey species. Scale worms, such as *Harmothoe* species, ragworms (*Platynereis*, *Nereis*), paddleworms (Phyllodocidae) and several genera of large, sedentary tube-builders such as *Neoamphitrite* and *Nicolea*, are not uncommon in kelp holdfasts in low-shore pools, and are not too difficult to identify. However, the polychaete faunas of *Corallina officinalis* and the red algal fringe present rather a challenge for the seashore ecologist. Quantitative studies of these habitats often reveal polychaete worms to be a significant, if not dominant, fraction of their communities, in terms of numbers of both individuals and species (e.g. Bussell *et al.* 2007). These are all small worms; some might be juveniles recently settled from the plankton and, as with many small gastropods, perhaps only transitory inhabitants, but many others will belong to the Syllidae, a family of very small worms, many no more than 2–3 mm long at maturity. Species of numerous syllid genera may be found in rock-pool algal turfs (e.g. *Odontosyllis*, *Sphaerosyllis*, *Brania*) and their actual diversity is probably under-recorded for most shores; they require patience, and a stereomicroscope, to identify them to species level.

There is no comprehensive account of the polychaetes of shallow north-west European coastal habitats. Several Linnean Society Synopses provide keys and descriptions to species of a few selected orders (Chambers & Muir 1997; George & Hartmann-Schröder 1985; Pleijel & Dales 1991; Westheide 2008), and a guide to the Syllidae of the British Isles is provided by San Martin & Worsfold (2015). Revised keys to the most frequently occurring families, genera and species of shallow-water polychaetes of the region have been presented by Knight-Jones *et al.* (2017), together with

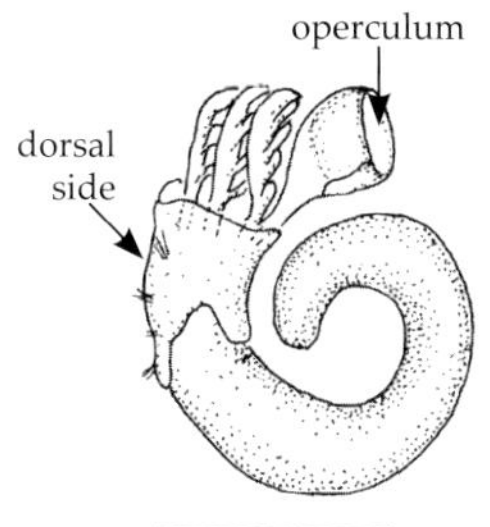

J.1 *Neodexiospira pseudocorrugata* removed from tube to show operculum and fused dorsal thoracic lobes. Scale bar: 0.5 mm.

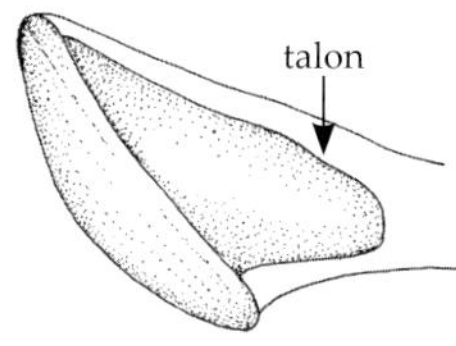

J.2 *Spirorbis corallinae* operculum, showing talon on side of stalk. Scale bar: 0.1 mm.

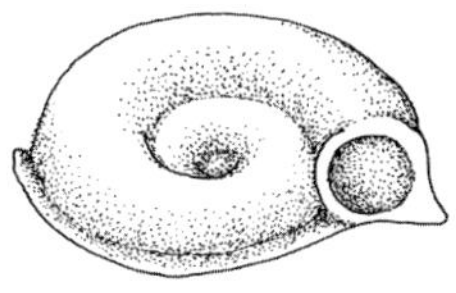

J.3 *Spirorbis spirorbis* tube. Scale bar: 1 mm.

an extensive bibliography of taxonomic works helpful in the identification of particular families and genera.

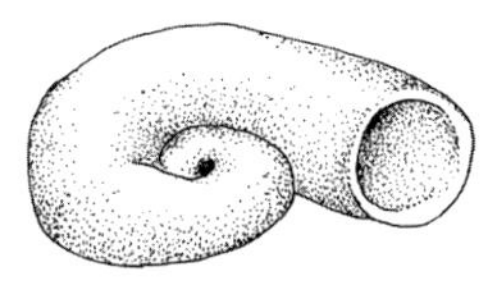

J.4 *Spirorbis rupestris* tube. Scale bar: 1 mm.

J.5 *Spirorbis corallinae* tube. Scale bar: 1 mm.

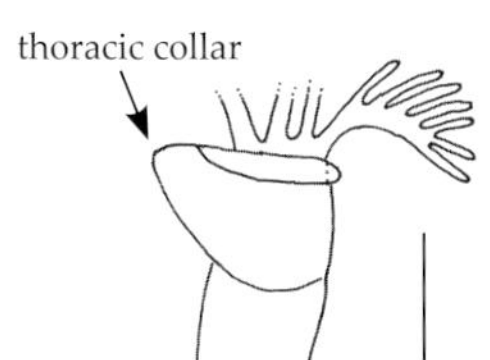

J.6 *Spirorbis corallinae*, showing thoracic collar. Scale bar: 0.5 mm.

1. Worm living in a white, spiralled calcareous tube, securely cemented to the substratum. Body with a crown of tentacles at the front end, one of which is modified as an operculum which closes the mouth of the tube (J.1) **2**

– Worm free living; or in a tube of sand or silt bonded with mucus. No operculum **7**

2. Tube with a clockwise spiral **3**

– Tube with an anticlockwise spiral **6**

3. Operculum with a well-developed plate (talon) on one side of its stalk (J.2) **5**

– Operculum without, or with only minimal development of, a talon **4**

4. Tube often with a flat rim where it meets substratum; body of worm pale green-brown (J.3). Characteristically on *Fucus serratus*, less frequently on *F. vesiculosus* or other large, lower-shore brown algae *Spirorbis spirorbis*

– Tube without a rim; body of worm red (J.4). Lower shore, on hard substrata, associated with the purple, encrusting coralline alga *Phymatolithon* *Spirorbis rupestris*

5. Tube small (less than 1.5 mm diameter (J.5)), final whorl typically overlapped onto previous ones. Collar of thorax asymmetrical (J.6), but not lobed. Intertidal and shallow subtidal; south and west coasts only. Almost exclusively on *Corallina officinalis* *Spirorbis corallinae*

– Tube diameter often greater than 1.5 mm, sometimes with flat rim, last whorl not overlapping rest. Collar of

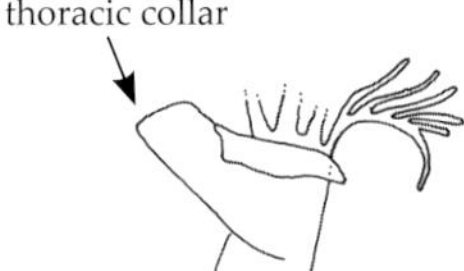

J.7 *Spirorbis inornatus*, showing thoracic collar. Scale bar: 0.5 mm.

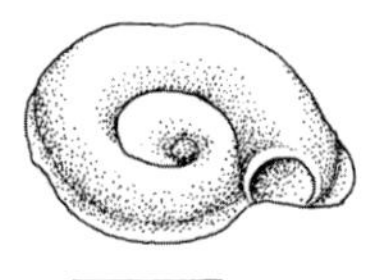

J.8 *Spirorbis inornatus* tube. Scale bar: 1 mm.

J.9 *Janua pagenstecheri*, removed from tube to show two unfused thoracic lobes. Scale bar: 0.5 mm.

J.10 *Janua pagenstecheri* tube. Scale bar: 1 mm.

thorax asymmetrical, with pronounced lobe on one side (J.7–J.8). Lower shore and shallow subtidal. On a range of algae: *Laminaria*, *Himanthalia* (underside of holdfast buttons), *Chondrus*, *Mastocarpus* and *Furcellaria* *Spirorbis inornatus*

6. Thoracic collar with two, unfused, dorsal lobes (J.9). Tube with ridges along its length, a middle ridge usually well marked in juvenile stages, up to three ridges in later growth, sometimes poorly developed (J.10). Intertidal and subtidal; widespread, common. On various substrata, including a range of algae *Janua pagenstecheri*

– Folds of thoracic collar fused dorsally (J.1). Three well-marked ridges along tube. Lower shore and subtidal, on *Saccharina latissima*, *Cystoseira* and small red algae. South-west coasts only *Neodexiospira pseudocorrugata*

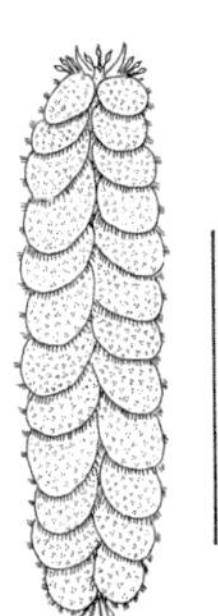

J.11 *Lepidonotus squamatus*. Scale bar: 2 cm.

7. Rather short, broad, oval worms; up to 5 cm long. Upper surface of body covered with overlapping series of flat scales (J.11), easily lost by careless handling **8**

– Not as described **11**

8. Twelve pairs of scales. Body light yellow to brown, scales spotted or patterned; up to 3 cm long (J.11). Mid- and low-shore, common, frequently in algal holdfasts *Lepidonotus squamatus*

– More than 12 pairs of scales **9**

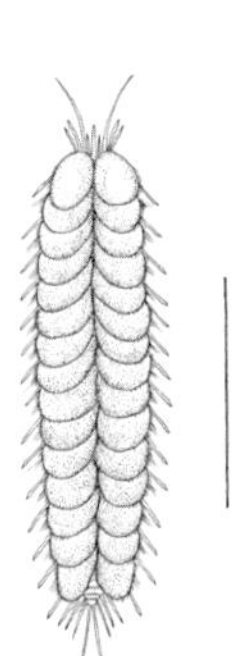

J.12 *Harmothoe impar*. Scale bar: 2 cm.

9. Tail of worm – up to ten segments – not covered by scales, bearing two long cirri at tip (J.14–J.15) **10**

– Body of worm completely covered by scales, each with granular surface and fringe of fine hairs; grey, green or brownish, often spotted or flecked. Up to 3 cm long (J.12). Intertidal and shallow sublittoral, widespread and common. Often amongst algae, and in holdfasts *Harmothoe impar*

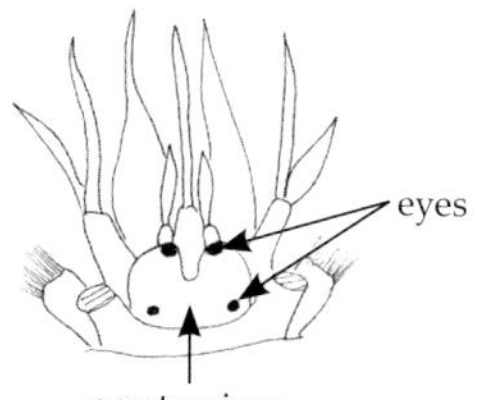

J.13 *Harmothoe imbricata*, detail of head showing position of eyes on prostomium.

10. First pair of eyes close to front of prostomium (J.13). Scales rough surfaced, covered with small papillae, colour variable, blue, grey or brown, to red, purple or black, often patterned. Up to 7 cm long (J.14). Intertidal and sublittoral, frequently in algal holdfasts *Harmothoe imbricata*

– First pair of eyes towards middle of prostomium, closer to second pair. Scales grey, brown or reddish, with a clear central area. Up to 7 cm long (J.15). Lower shore, widespread and common, in *Laminaria* holdfasts *Harmothoe extenuata*

11. Front end of worm with a tangled mass of thread-like tentacles; ***or*** a more regular fan of stiff, feathery ones. Living in tube: leathery, with sand, gravel and shell attached; ***or*** formed from mucus-bound sand and silt **12**

– Front end of worm with various paired appendages but lacking a tentacle crown, head usually distinct **14**

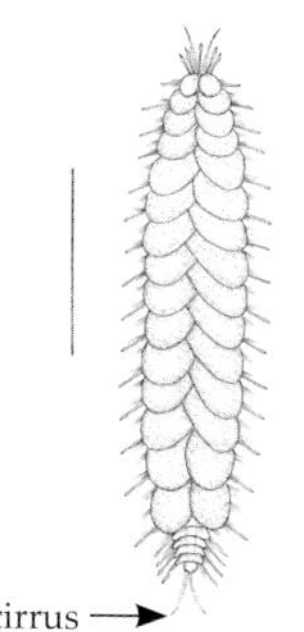

J.14 *Harmothoe imbricata*. Scale bar: 2 cm.

12. Front end of worm with mass of thread-like tentacles; two pairs of red, branching gills behind tentacles; thorax of 15 segments; up to 6 cm long (J.16). Tube of sand and detritus, attached to algae, hydroids, sponges and bryozoan turf *Nicolea venustula*

Nicolea zostericola is similar but smaller, to 2 cm, with 17–18 thoracic segments and long-stalked gills. Also similar, but much larger (to 30 cm long), are *Neoamphitrite edwardsii* and *N. figulus*, both of which have three pairs of gills behind the tentacles.

– Front end of worm with stiff, regular tentacle fan. Tube tough, leathery **13**

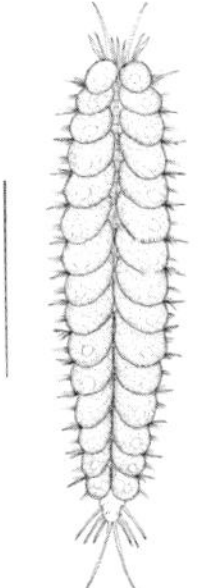

J.15 *Harmothoe extenuata*. Scale bar: 2 cm.

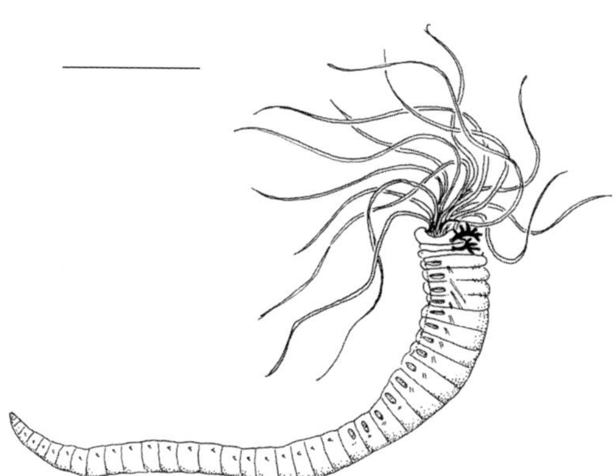

J.16 *Nicolea venustula*: left-hand view, with paired gills shown black. Scale bar: 5 mm.

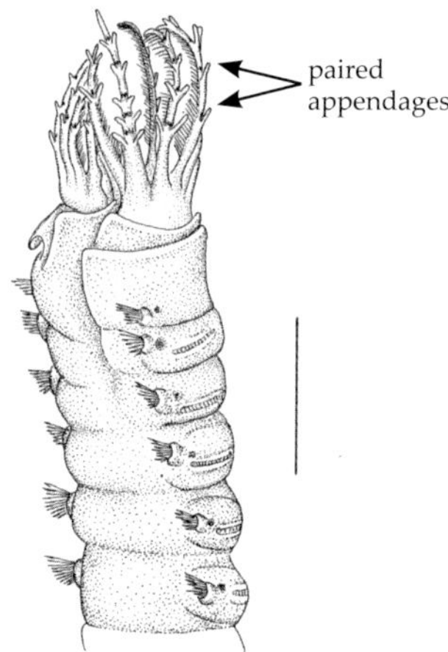

J.17 *Branchiomma bombyx*, showing paired tentacle appendages. Scale bar: 1 mm.

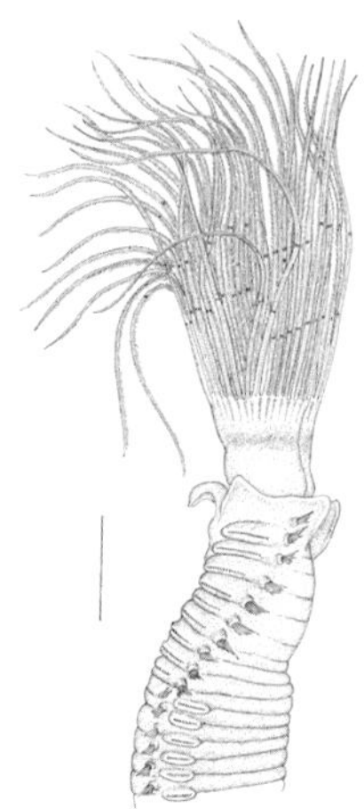

J.18 *Megalomma vesiculosum*, showing eyespots on tentacles. Scale bar: 5 mm.

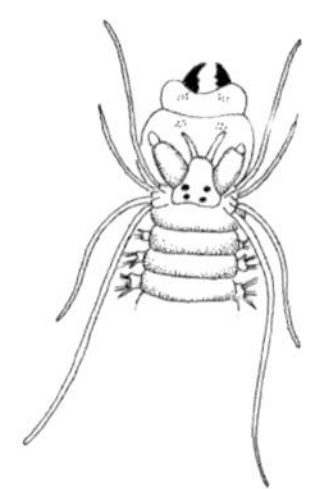

J.19 *Platynereis dumerilii*: head and anterior part of body, with proboscis everted.

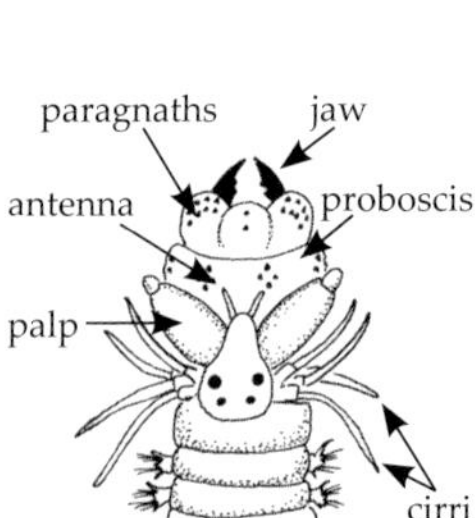

J.20 *Nereis pelagica*: head and anterior end of body, with proboscis everted.

13. Tentacles rather short and stout, with paired appendages. Up to 5 cm long (J.17). Low shore and shallow subtidal, often in kelp holdfasts *Branchiomma bombyx*

– Tentacles long and slender, without appendages, but with distinct eyespots. Up to 15 cm long (J.18). Low shore and shallow subtidal, south and west coasts. Often in kelp holdfasts *Megalomma vesiculosum*

14. Worms with well-developed heads bearing paired antennae, paired palps and several pairs of tentacle-like cirri; eversible proboscis (gentle pressure with finger, applied just behind the head of the worm, will cause the proboscis to evert) with large, powerful jaws (J.19) **15**

– Not as described **17**

15. Proboscis, when everted, with distinct groups or lines of small black teeth (paragnaths) (J.20); cirri rather short **16**

– Proboscis with very small, poorly developed paragnaths. Cirri very long, reaching back to at least body segment 10. Up to 6 cm long, slender (J.19). Lower shore and shallow subtidal, in kelp holdfasts, where it may secrete mucous tubes *Platynereis dumerilii*

16. Paragnaths in two small groups. Up to 12 cm long; colour variable, often greenish-bronze (J.20). Lower shore, amongst algae and in holdfasts *Nereis pelagica*

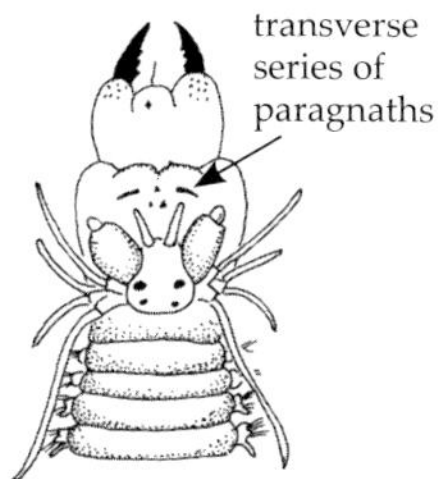

J.21 *Perinereis cultrifera*: head and anterior end of body, with proboscis everted.

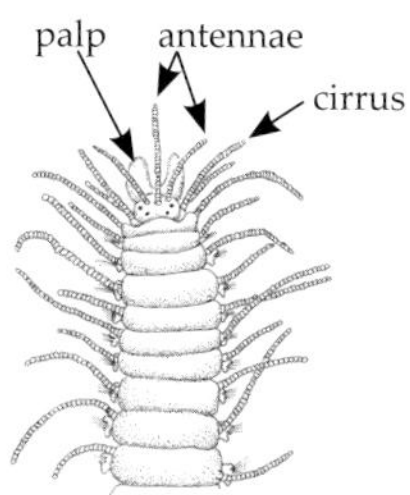

J.22 *Syllis* species: head and anterior end of body, showing palps clearly separated.

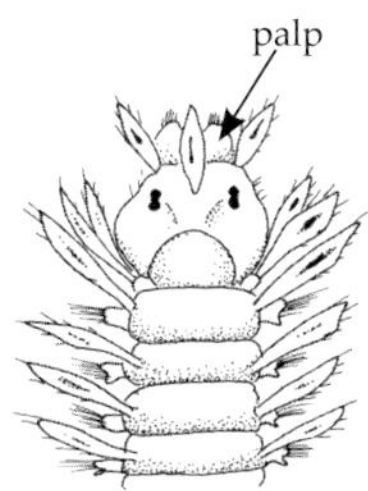

J.23 *Odontosyllis* species: head and anterior end of body, showing palps fused at base.

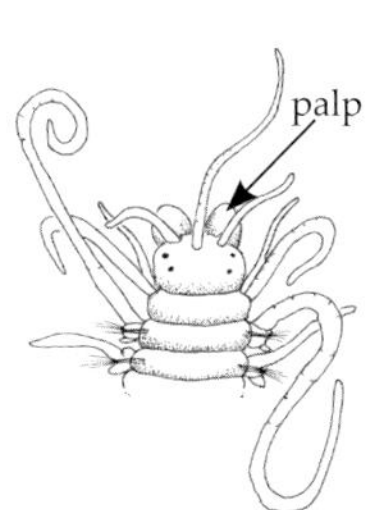

J.24 *Eusyllis* species: head and anterior end of body, showing palps fused for most of length.

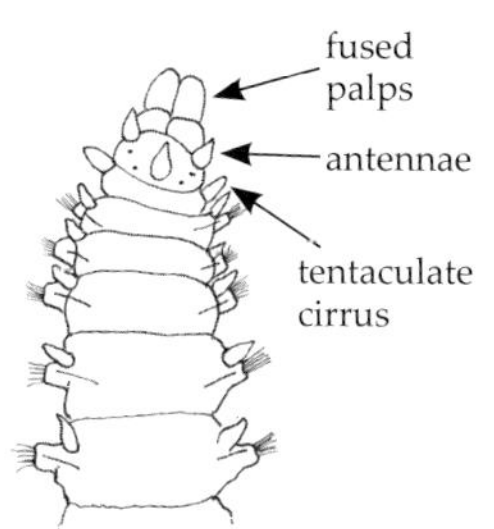

J.25 *Sphaerosyllis* species: head and anterior end of body, showing palps fused for most of length.

– Paragnaths in distinct transverse series. Up to 25 cm long; greenish bronze, with red tints (J.21). Low shore, amongst algae and in holdfasts *Perinereis cultrifera*

17. Head of worm bearing three antennae, two short palps and one or two pairs of tentacle-like cirri (J.22) **18**

– Head of worm with two, four or five antennae **20**

18. Head with palps clearly separated. Each body segment with a pair of elongate, jointed cirri. Delicate, slender worms, up to 3 cm long (J.22). Often common in algal holdfasts *Syllis* species

– Head with palps fused at base (J.23), or for most of length (J.24); cirri on body segments smooth or only indistinctly jointed **19**

19. Head with palps fused at base. Small, fragile worms, often common in algal turfs and kelp holdfasts (J.23, J.24) Numerous species, in many genera, including *Odontosyllis* and *Eusyllis*

– Head with palps fused for half to whole of length (J.25). Numerous species of small (less than 2 cm) or very small (2–4 mm) worms, often common in algal turfs and kelp holdfasts Many genera, including *Brania*, *Salvatoria* and *Sphaerosyllis*

parapodia
(sing. parapodium) paired appendages borne on each body segment; usually bilobed, unilobed in some genera, bearing chaetae, gills or tentacle-like processes – cirri – in many species

20. Head of worm with one or two pairs of antennae and two or three pairs of long tentacle-like cirri (J.27). Body parapodia each bearing a long cirrus (J.28) **21**

– Head of worm with four or five antennae and four pairs of rather short tentacle-like cirri. Parapodia with large, paddle-like cirri (J.26) Phyllodocidae (paddleworms)

These large, distinctive worms, often more than 15 cm in length, coloured green, yellow or brown, in spots and bands, can be found low on the shore, beneath stones, in crevices and kelp holdfasts in large pools. Pleijel & Dales (1991) provide keys to north-west European species.

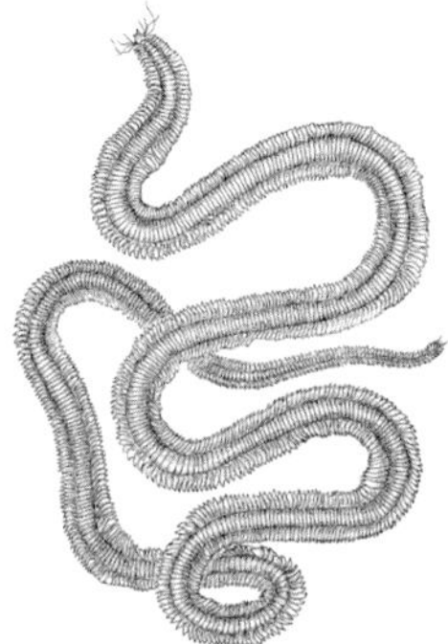

J.26 A paddle worm, *Phyllodoce* species

21. Head with two pairs of antennae and two pairs of tentacle-like cirri. Up to 8 cm long; yellow, brown or reddish (J.27). Lower shore, often in *Laminaria* holdfasts *Kefersteinia cirrata*

– Head with one pair of antennae and three pairs of tentacle-like cirri. Up to 3 cm long, yellowish with dark banding (J.28). Lower shore and subtidal, often in *Laminaria* holdfasts *Castalia punctata*

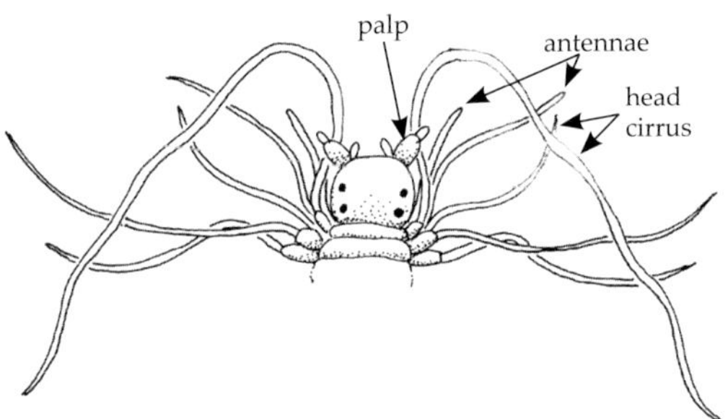

J.27 *Kefersteinia cirrata*: head and anterior end of body.

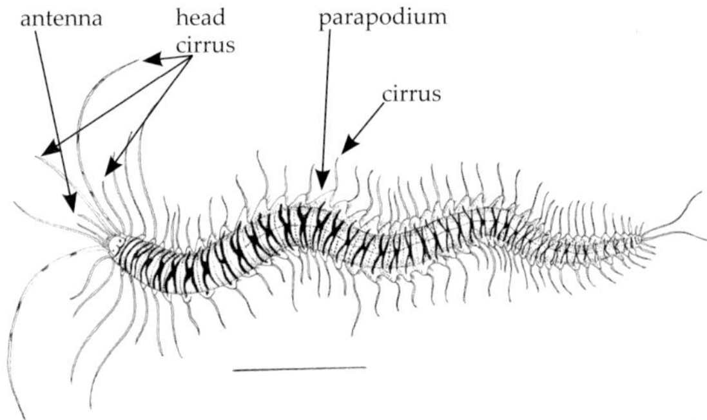

J.28 *Castalia punctata*. Scale bar: 2 cm.

Key K Heterobranchia

The Heterobranchia is possibly the most diverse subclass of the Gastropoda. It encompasses several families of tiny shelled snails, presently unassigned to any higher taxonomic rank, several orders characterised by the presence of a small or vestigial shell that does not accommodate the entire body of the animal, and the large order Nudibranchia, in which the shell is entirely lacking. The nudibranchs, together with the orders Pleurobranchida, Cephalaspidea, Runcinida and Aplysiida, and the superorder Sacoglossa, are often still loosely referred to as 'sea slugs' or opisthobranchs, although neither term has any taxonomic validity. These soft-bodied gastropods are most easily identified whilst still alive, because colours and structures such as gills, which are only ever everted in the living animal are important characters. The British 'opisthobranch' species are described in two Ray Society monographs (Thompson 1976; Thompson & Brown 1984), while synoptic accounts, together with identification keys, are provided by Thompson (1988). All three works, though still useful sources, are currently out of print, and the taxonomy and classifications they employ are now completely outdated. Picton & Morrow (1994) published a useful photographic guide to nudibranchs, and a broader selection of heterobranchs can be found in their online *Marine Life Encyclopaedia* (Picton & Morrow 2016). Keys for the identification of the most common heterobranch species are presented in Hayward & Ryland (2017).

1. Body slender, posteriorly tapered, the head with one or two pairs of tentacular processes. A pair of broad dorsal flaps (parapodial lobes) run the length of the body, converging close to the posterior end (K.1–K.2) **2**
– Head with or without tentacular processes, dorsal surface smooth, *or* tuberculate, *or* with various processes, but without paired parapodial lobes **3**

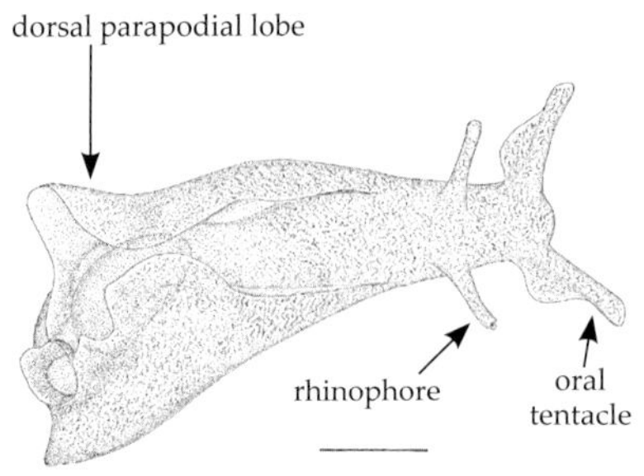

K.1 *Aplysia punctata*. Scale bar: 10 mm. (From Hayward & Ryland 2017)

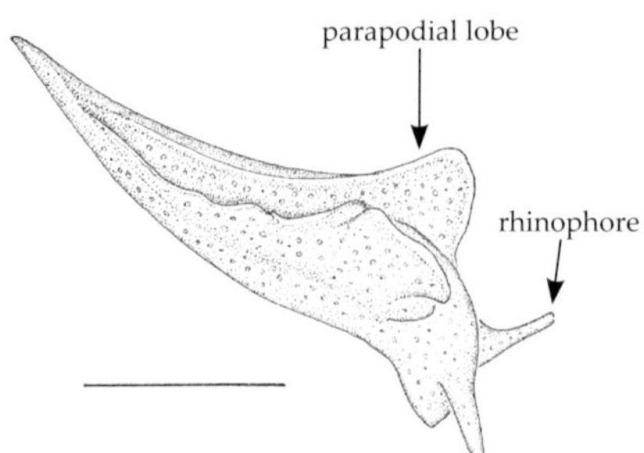

K.2 *Elysia viridis*. Scale bar: 5 mm. (From Hayward & Ryland 2017.)

rhinophores
sensory appendages borne by many species of heterobranch

2. Head with a pair of enrolled oral tentacles anteriorly, and behind them paired, enrolled rhinophores. A very small, translucent, shell rudiment posteriorly, close to the convergence of the parapodia (K.1). Herbivorous. Red, dark green, brown or purplish black, depending on diet. Up to 30 cm long. Juveniles may occur on small red or green algae in low-shore pools *Aplysia punctata*

– Head with a pair of prominent enrolled rhinophores but no oral tentacles; a pair of shorter propodial tentacles (K.2) arise from the foot, below the head. No shell rudiment. Up to 35 mm long. Bright green, brown or red, with scattered flecks of brilliant blue, red and green. On small red and green algae, particularly species of *Codium* *Elysia viridis*

3. Small, elongate, smooth bodied, 4–8 mm long; with or without head appendages, but with no other dorsal processes **4**

– Upper surface of body coarsely tuberculate, ***or*** with rows of finger-like, clubbed or leaf – like processes, and often a circlet of pinnate gills posteriorly **6**

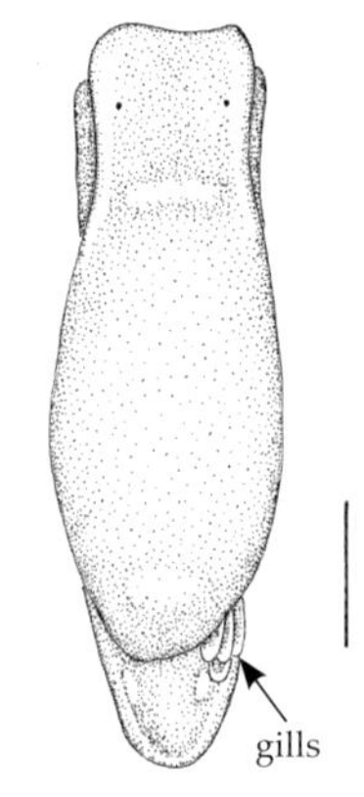

K.3 *Runcina coronata*. Scale bar: 1 mm.

4. Smooth, elongate; head rather square, lacking tentacles. Conspicuous gills projecting posteriorly on right side. Dark brown with pale patches on head and tail; up to 6 mm long (K.3). Often in rock pools, feeding on small algae, particularly *Codium* *Runcina coronata*

– Head with or without tentacles; body tapered posteriorly, but without gills **5**

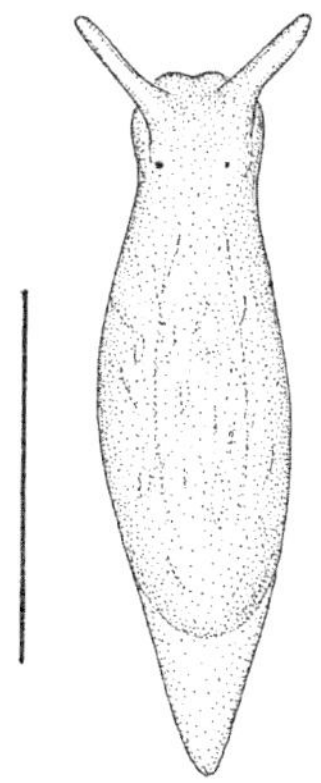

K.4 *Limapontia senestra.* Scale bar: 1 mm.

5. Head with a pair of slender rhinophoral tentacles. Up to 6 mm long, olive brown to black, typically marbled, with lighter patches around eyes (K.4). On *Cladophora* and *Ulva* *Limapontia senestra*
– Head lacking tentacles, but with two short longitudinal ridges. Up to 4 mm long (rarely, 8 mm), dark brown to black (K.5). On *Cladophora* and *Enteromorpha* *Limapontia capitata*

6. Body oval, domed, with coarsely tuberculate dorsal surface; two short, club-ended tentacles project from beneath anterior edge in active animals, but other processes absent. Dark olive green to black, up to 12 mm long (K.6). Coasts of Devon and Cornwall; upper shore, emerging at low tide to feed around pool edges, retreating upshore with incoming tide *Onchidella celtica*
– Dorsal surface smooth or tuberculate, but always with a variety of processes – rhinophores, oral tentacles, rows of cerata, or posterior gill circlets **7**

ceras (pl. cerata)
extensions of the body wall, finger- or leaf-like, branching or knobbly, enclosing lobes of the digestive gland; may have conspicuous terminal sac containing stinging cells derived from hydroid prey, and may also function as gills

7. Dorsal surface with rows of cerata: flattened and leaf-like, cylindrical or club-shaped with rings of large tubercles **8**
– Without rows of cerata, but with prominent rhinophores, cephalic tentacles or gills **13**

8. Body slender, narrowly tapered posteriorly, with transverse rows of delicate leaf-like or cylindrical cerata. Head rounded, bearing a pair of simple, enrolled rhinophores. Anus dorsal, anterior, visible behind the rhinophores **9**
– Not as described. Head typically with a pair of long cephalic tentacles anterior to the rhinophores, which may have distinct basal sheaths (K.9–K.10) **10**

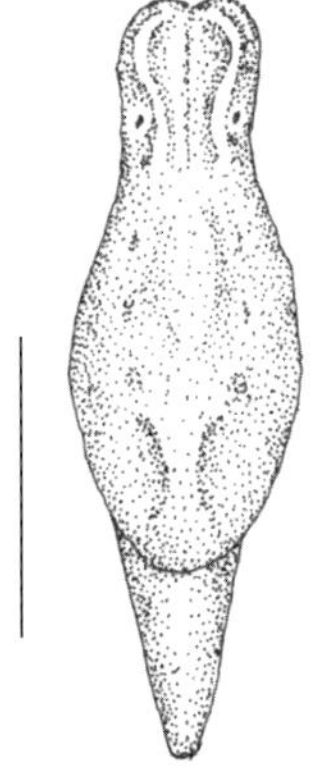

K.5 *Limapontia capitata.* Scale bar: 1 mm.

K.6 *Onchidella celtica.* (Photo: J.S. Ryland)

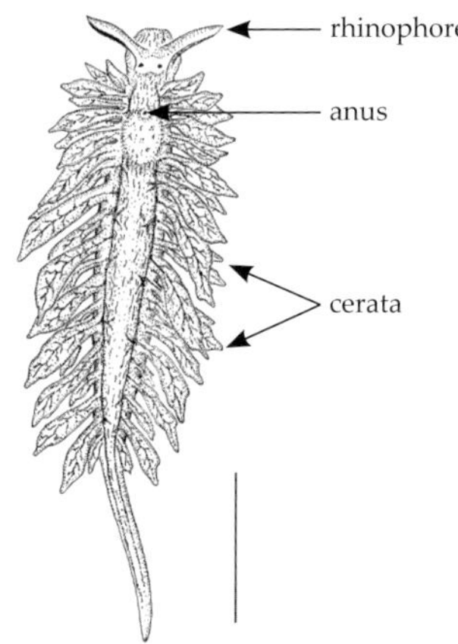

K.7 *Hermaea bifida*. Scale bar: 5 mm.

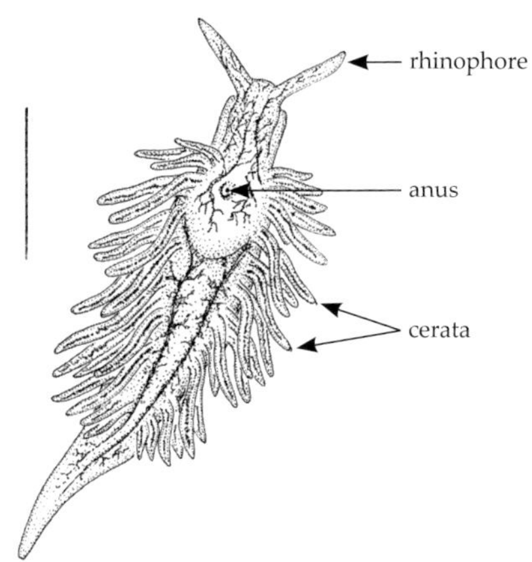

K.8 *Placida dendritica*. Scale bar: 2 mm.

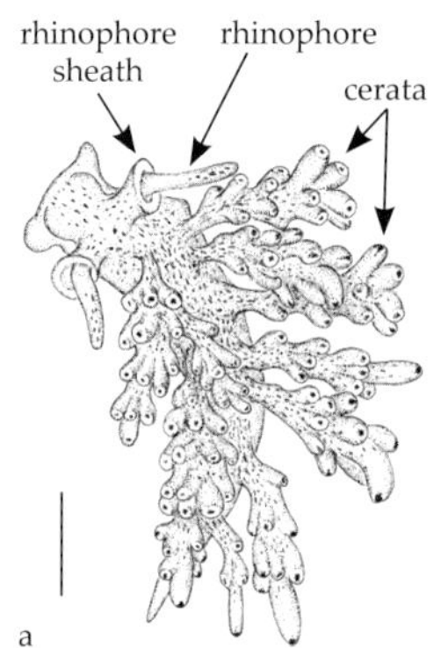

9. Body white, tinted rose-pink at front. Cerata leaf-like. Up to 2 cm long (K.7). Lower shore, on small red algae *Hermaea bifida*

– Body greenish white. Cerata slender, cylindrical, rounded at tip. Up to 11 mm long (K.8). Lower shore, on *Codium* and *Bryopsis* *Placida dendritica*

10. Rhinophores simple, finger-like, with trumpet-shaped sheaths. About eight pairs of club-shaped, knobbly cerata. Body yellow or white, with red or purple spots and blotches. Up to 12 mm long (K.9a). Common on a number of hydroids, including species of *Sertularia*, *Obelia* and *Tubularia* *Doto coronata*

– Rhinophores smooth or ridged, without sheaths **11**

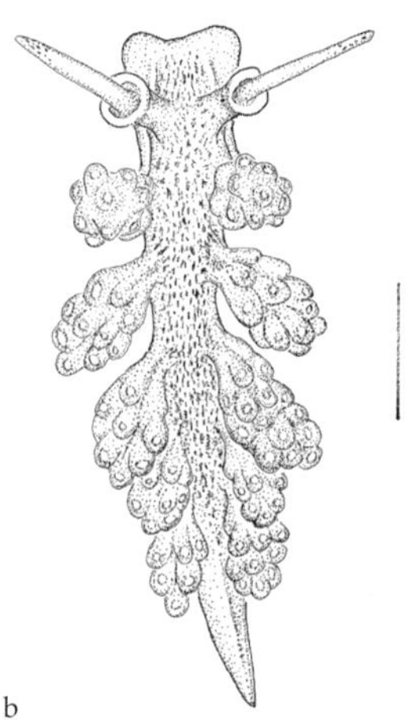

K.9 *Doto* species. **a** *Doto coronata* (scale bar: 1 mm); **b** *Doto lemchei* (scale bar: 2 mm); **c** *Doto fragilis*. Note flared rhinophore sheaths. (Photo: J.S. Ryland)

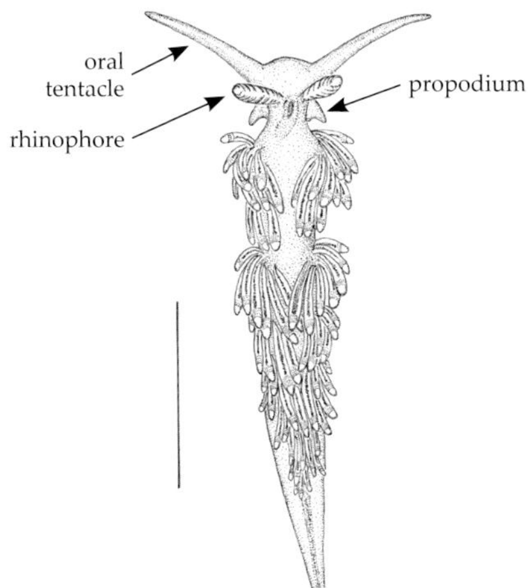

K.10 *Facelina auriculata*. Scale bar: 10 mm.

K.11 *Flabellina lineata*. Scale bar: 2 mm.

11. Rhinophores with transverse ridges. Cerata in dense clusters on each side of the midline of the body. Body white, tinged with pink. Up to 5 cm long (K.10). Feeds on a broad range of hydroids *Facelina auriculata*

– Rhinophores smooth, or with faint knobs; without marked transverse ridges **12**

12. Anterior edge of foot (the propodium) with paired propodial tentacles; rhinophores with wrinkled or faintly tuberculate surface. Cerata in distinct clusters. Up to 4 cm long. Body white or violet, sometimes with white lines or blotches; cerata red or orange, often with white tips (K.11). Predators of hydroids *Flabellina* species

– With oral tentacles and rhinophores only; no propodial tentacles. Cerata stout or slender, often swollen or club shaped, in transverse rows; typically colour banded. Length 1–3 cm Tergipedidae

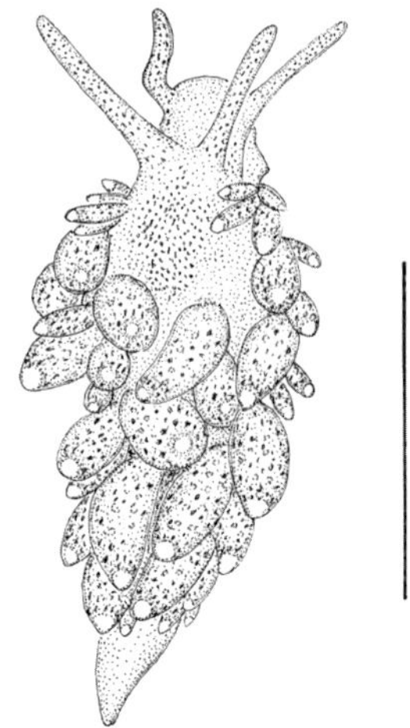

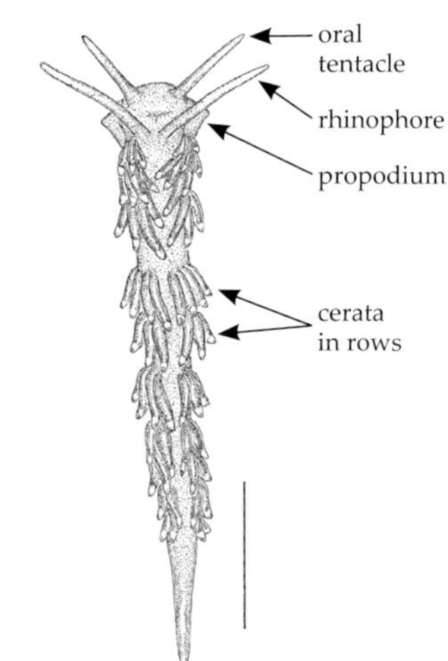

K.12 *Eubranchus pallidus*. Scale bar: 10 mm.

K.13 *Diaphoreolis viridis*. Scale bar: 5 mm.

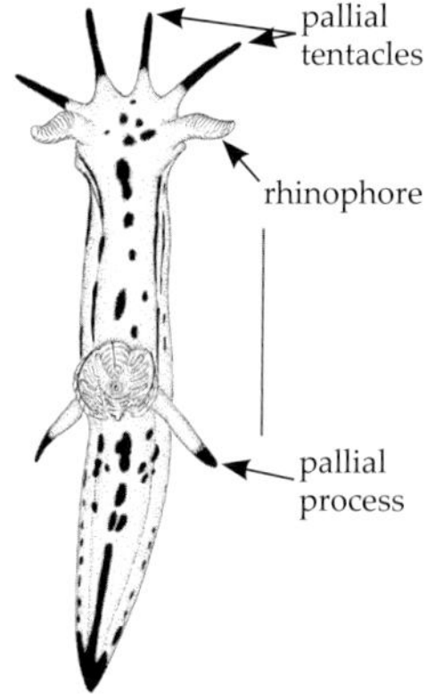

K.14 *Polycera quadrilineata*. Scale bar: 10 mm.

cnidosac
a sac at the tip of the ceras in which stinging cells (cnidocytes) derived from hydroid prey are stored, and employed as defence

A small family of nudibranchs with 12 species presently recorded for north-west European coasts. All are predators of hydroids. *Eubranchus pallidus* (K.12), around 2 cm long, has a greyish ground colour with brownish red, yellow and white blotches on the body and club-shaped cerata; rhinophores and oral tentacles are banded brown, with dense white spots towards their tips. *Diaphoreolis viridis* (K.13) is slender bodied, up to around 15 mm long, pale translucent yellow, with tapered, green cerata, and conspicuous white cnidosacs.

13. Body elongate, slender, smooth, anterior edge with 4–6 (rarely, more) slender, finger-like processes; posteriorly two similar processes (pallial tentacles), with a circlet of pinnate gills between. Body white, with yellow or orange blotches and a few black streaks. Up to 4 cm long (K.14). A predator of the bryozoan *Membranipora*, on kelp fronds *Polycera quadrilineata*

– Dorsal surface rough textured, covered with coarse papillae of varying sizes; posteriorly, gills arranged in a circle around the anus **14**

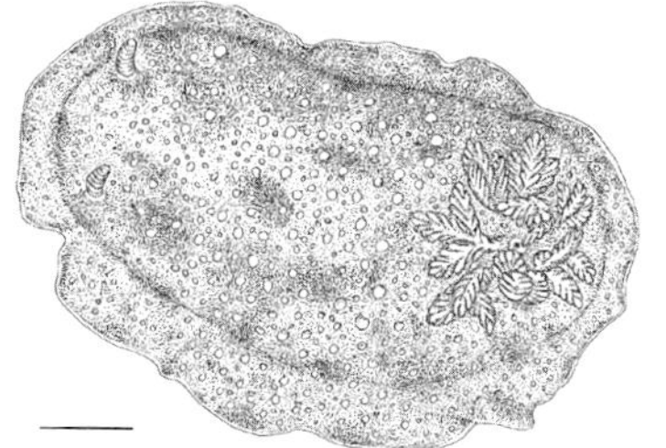

K.15 *Doris pseudoargus*. Scale bar: 10 mm.

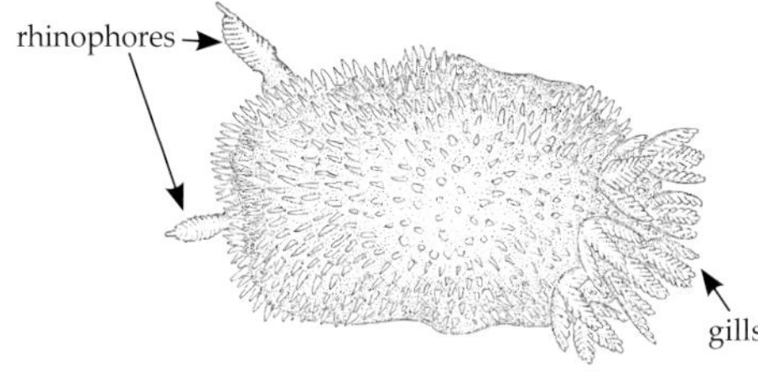

K.16 *Acanthodoris pilosa*. Scale bar: 10 mm.

14. When disturbed, animal retracts gills simultaneously into a deep pocket. Colour variable, but always an irregular patchwork of yellow, brown, green, white or pinkish blotches. Large, solid and fleshy; commonly 6–8 cm long, but up to 12 cm (K.15). Feeds on sponges, particularly *Halichondria*, often on kelp holdfasts *Doris pseudoargus*

– When disturbed, animal retracts gills individually. Usually less than 4 cm long **15**

15. Rhinophores with short, frilled sheaths. Mantle papillae elongate, conical, soft. Commonly up to 3 cm long. Uniformly white, grey or brown, of varying shades (K.16). A predator of encrusting bryozoans, particularly *Flustrellidra hispida* and *Alcyonidium* species *Acanthodoris pilosa*

– Rhinophores without sheaths. Mantle papillae short, rounded, stiff **16**

16. Mantle papillae relatively large, club shaped. Gill circlet large, with up to 29 gills. Up to 4 cm long (K.17). Adults prey on barnacles and encrusting, calcified bryozoans, juveniles on barnacles *Onchidoris bilamellata*

– Mantle papillae small, spiny. Gill circlet with about 12 gills. Up to 9 mm long (K.18). A predator of encrusting calcified bryozoans *Onchidoris muricata*

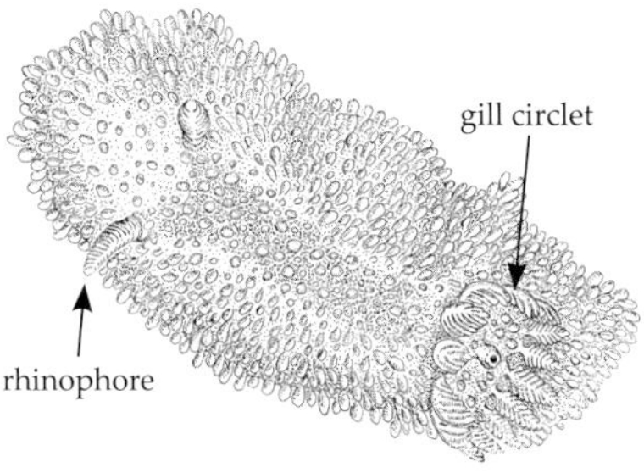

K.17 *Onchidoris bilamellata*. Scale bar: 10 mm.

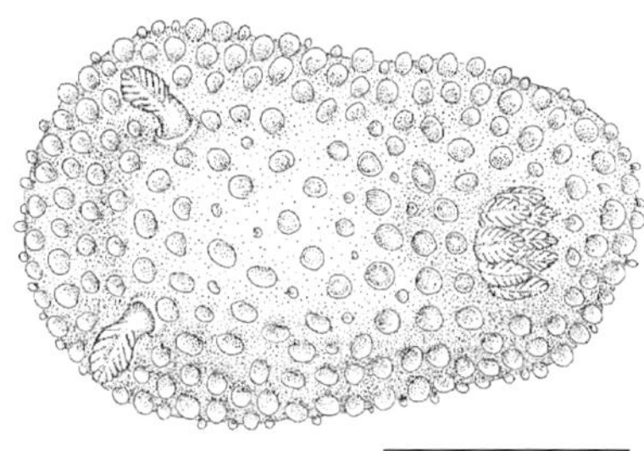

K.18 *Onchidoris muricata*. Scale bar: 5 mm.

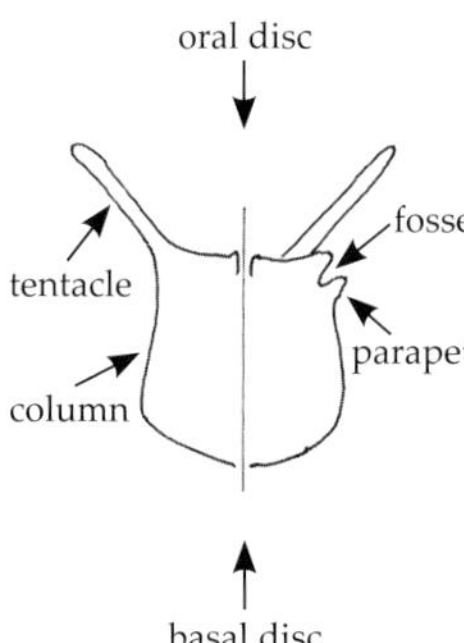

L.1 Gross morphological features of an anemone, without (left) and with (right) a parapet.

Key L Sea anemones

Several species of the cnidarian order Actiniaria, the sea anemones, are common and conspicuous components of rocky intertidal habitats and, with the exception of the Beadlet Anemone *Actinia equina*, are dependent on standing pools for their survival through tidal emersion. A few more may be present in cryptic habitats in low-shore pools. They are radially symmetrical animals, with bodies approximating to a contractile cylinder. The base, except in modified burrowing species, forms a flat basal disc (L.1), which secures the animal to the substratum; the opposite end is the oral disc, with a central mouth, surrounded by tentacles, the number and arrangement of which varies between species and genera. The region between basal and oral discs is referred to as the column; it may be smooth surfaced, or bear rows of warts (verrucae) or adhesive suckers. In some families the distal end of the column is folded to form a projecting rim, the parapet, below the oral disc, from which it is separated by a distinct gutter, termed the fosse.

There are some practical difficulties in identifying sea anemones. The more common species, such as *Actinia equina*, *Urticina felina* and *Sagartia elegans*, are easily recognised by their distinctive colour patterns and gross morphologies, but for others a closer examination of both the expanded disc and the morphology of the column is necessary. They are best identified in the field: most species contract swiftly on disturbance and are slow to open again; it is generally futile to try and remove them for later examination. Such features as number and arrangement of tentacles, colour and patterning of disc and tentacles should be recorded before the animals are disturbed (a digital camera is an ideal tool), in order to look for diagnostic features of the column. Colour images of all British species of Actiniaria are illustrated, in expanded oral disc view, by Wood (2013), who also provides notes on their biology, ecology and distribution. Dichotomous keys for the identification of the most common species on north-west European coasts are provided by Hayward & Ryland (2017).

L.2 The distinctive tentacles of two individuals of the Snakelocks Anemone *Anemonia viridis*, in a small mid-shore pool.

acrorhagi
prominent tubercles armed with clusters of stinging cells, employed in aggressive spatial competition

1. Tentacles always apparent, up to 200 in total, usually obscuring the column, may contract slightly on disturbance, but are not completely retractile. Disc diameter up to 5 cm. Brownish, grey green to bright green, tentacles usually light green, and often tipped with purple (L.2). In mid- to low-shore pools, more common on exposed shores; south and west coasts of Britain and Ireland, absent from North Sea coasts Snakelocks Anemone *Anemonia viridis*

– Tentacles retractile, column and disc always visible **2**

2. Rim of column forming a projecting parapet, encircling a fosse **3**

   Note that the Beadlet Anemone *Actinia equina* (see couplet 3, L.3) may be recognised at this point by its distinctive burgundy red colour, and as the only anemone consistently found out of water during tidal emersion, contracted to a glistening jelly blob.

– Column with neither parapet nor fosse **8**

3. Column smooth, disc bordered by a ring of blue acrorhagi, situated within the fosse. Up to 5 cm height and diameter,

L.3 The Beadlet Anemone *Actinia equina*: typically, in a damp crevice, covered with a film of water during low tide.

L.4 A group of Dahlia Anemones *Urticina felina*, typically individually patterned. (Photo: J.S. Porter)

with up to 200 tentacles; unicolorous, dull red, green or brown (L.3). Intertidal, from MHWN to ELWS, typically on damp, shaded rock surfaces, retreating into pools in winter. Conspicuous, and common on all rocky coasts Beadlet Anemone *Actinia equina*

The Strawberry Anemone *Actinia fragacea* is similar but larger (up to 10 cm), and the red column is flecked with green. It has a similar distribution to *A. equina* but is less common.

– Column smooth or with wart-like verrucae, no acrorhagi around disc **4**

4. Tentacles arranged in concentric cycles of ten – the innermost (around the mouth) 10, the second 10, the third 20, etc., up to a total of about 160; all are rather short and thick. Column with numerous verrucae, typically with adherent sand, shell fragments or other debris. Tentacle span up to 20 cm (L.4). Colour variable: grey, pale orange to red, often in patches, with colour-banded tentacles. Usually in close groups, ranging to around MHWN in deepest, shaded pools Dahlia Anemone *Urticina felina*

– Tentacles in cycles of six (6, 6, 12, 24, etc.), or without regular arrangement **5**

5. Disc with lobed edge, bearing hundreds (>1,000) of short tentacles, imparting a fluffy appearance, the parapet visible as a prominent collar below. Up to 25 cm high, with tentacle span of 15 cm or more. White to orange,

L.5 A large individual of the Plumose Anemone *Metridium dianthus*, with tentacles fully extended. (Photo: J.S. Porter)

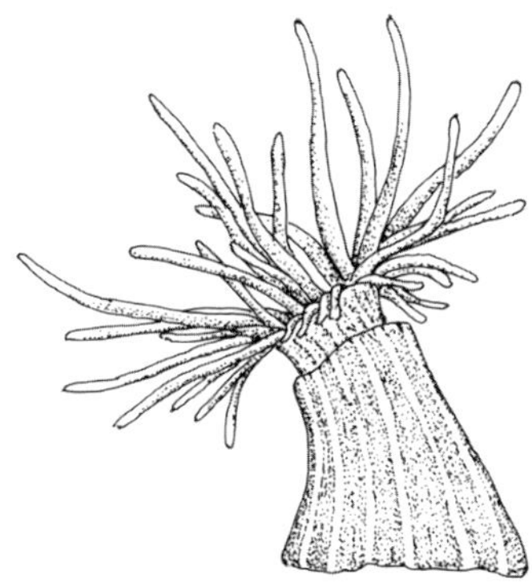

L.6 *Diadumene lineata*.

typically unicolorous (L.5). On all coasts, may occur in large, low-shore pools, forming dense clonal aggregations subtidally, on any hard substrata
Plumose Anemone *Metridium dianthus*

– Not as described; smaller, with fewer than 100, longer, tentacles, not 'plumose' **6**

6. Column smooth, without verrucae; brownish green, with vertical stripes of orange, yellow or white. Up to 100 irregularly arranged tentacles, with spread of 1–2 cm (L.6). Intertidal. An introduced species now found on all coasts in the region *Diadumene lineata*

– Column with vertical rows of verrucae. Variously coloured, but not striped **7**

7. Column up to 4 cm high, greyish pink, with vertical rows of verrucae, mostly slightly darker, except for six regularly spaced rows, which are conspicuously white. Up to 48 pale, grey-banded tentacles, with span of about 6 cm (L.7). Mostly intertidal, in pools, amongst *Corallina*

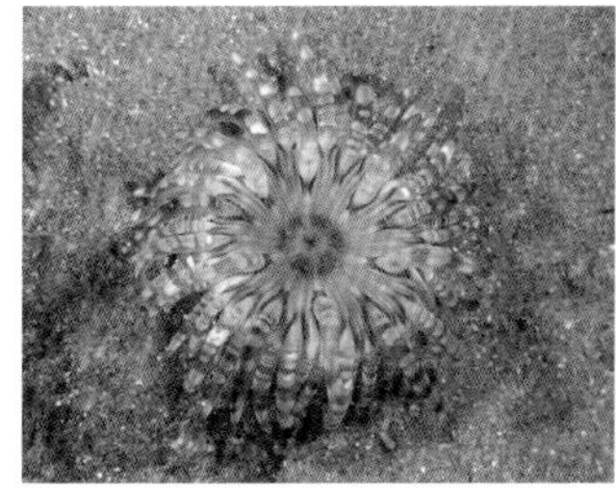

L.7 The Gem Anemone *Aulactinia verrucosa*: **a** contracted, showing the characteristic rows of white verrucae; **b** a small individual, expanded, showing banded tentacles; **c** a larger individual in disc view. (Photos: D.N. Huxtable/J.S. Ryland)

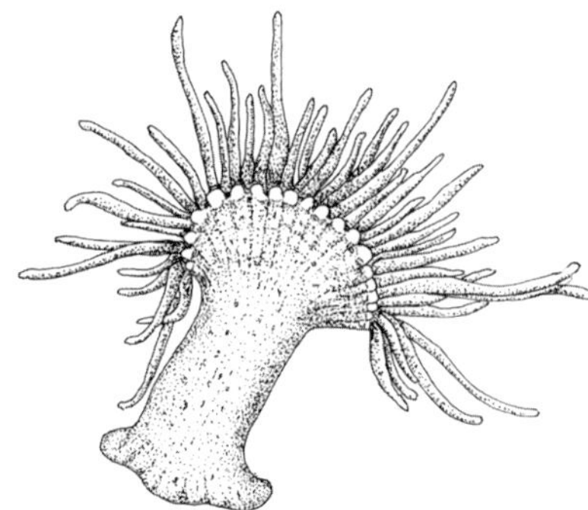

L.8 *Anthopleura ballii*: **a** showing column; **b** disc view. (Photo: D.N. Huxtable/J.S. Ryland)

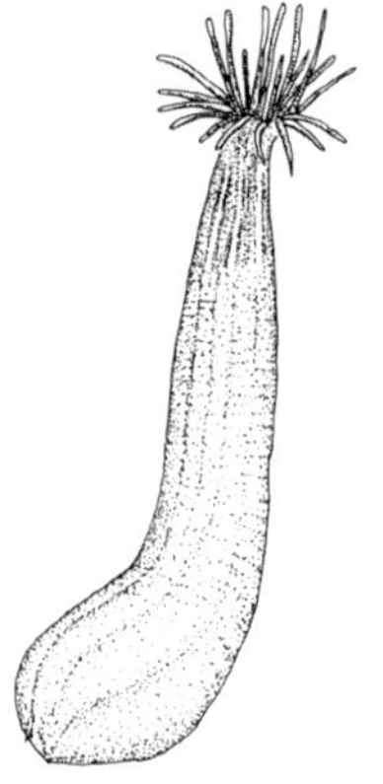

L.9 *Edwardsiella carnea.*

and in crevices. South and west coasts of England, Wales and Ireland, scarce south-west Scotland
Gem Anemone *Aulactinia verrucosa*

– Column up to 7 cm high, yellow to brown, flecked with white; small verrucae in vertical rows paler, each tipped with distinctive red spot; extending on to parapet, but no acrorhagi in fosse. Up to 96 tentacles, with span of about 10 cm (L.8). Intertidal, mid- and low-shore pools, to shallow sublittoral. South and west coasts of England, Wales and Ireland *Anthopleura ballii*

Anthopleura thallia is similar, but the tentacles are irregularly arranged rather than in cycles of six; verrucae tend to be darker than the column, often with adhering debris, and the fosse has a ring of pink or white acrorhagi. Intertidal, in pools from MTL to LWST. Rare: south-west England, Pembrokeshire, western Ireland.

8. Column a slender cylinder, 20 × 3–4 mm, rounded at proximal end, with no basal disc. Fewer than 36 tentacles. Translucent orange (L.9). Intertidal, from MTL, and shallow subtidal, in shaded rock crevices, often in large aggregations. South and west Britain, from Devon to the Western Isles *Edwardsiella carnea*

– Anemone with distinct, adhesive, basal disc, and more than 100 tentacles **9**

9. Column with conspicuous, light-coloured suckers, some of which may have adherent debris; smaller cinclides apparent towards top of column **10**

– Column without suckers, but with cinclides – seen as tiny spots – which may emit acontia **11**

acontia
thin filaments armed with stinging cells, ejected as a defensive response to disturbance

10. Disc broad, typically 4–5 cm diameter, with an average 500–700 small tentacles, in hexamerous (six-rayed) cycles; mottled light brown and blue grey, speckled with white.

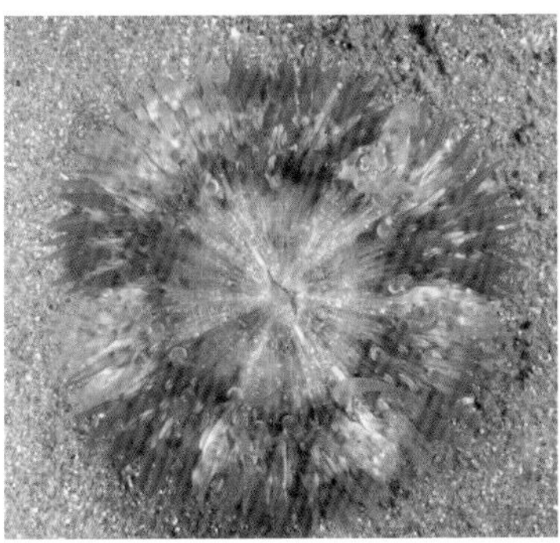

L.10 The Daisy Anemone *Cereus pedunculatus*; typically with column buried, the disc expanded at the sediment surface. (Photo: D.N. Huxtable)

Suckers appear as conspicuous white spots on the darker column, often with adherent debris. Cinclides, present towards top of column, readily emit acontia (L.10). Intertidal, up to MTL, and shallow subtidal. South and west coasts of Britain, north to Shetland, all Irish coasts Daisy Anemone *Cereus pedunculatus*

– Column tall, with broad base – 3 cm diameter – and flared distally, with up to 200 irregularly arranged tentacles. Suckers usually without adherent debris, cinclides as small spots towards top of column (L.11). Intertidal, often

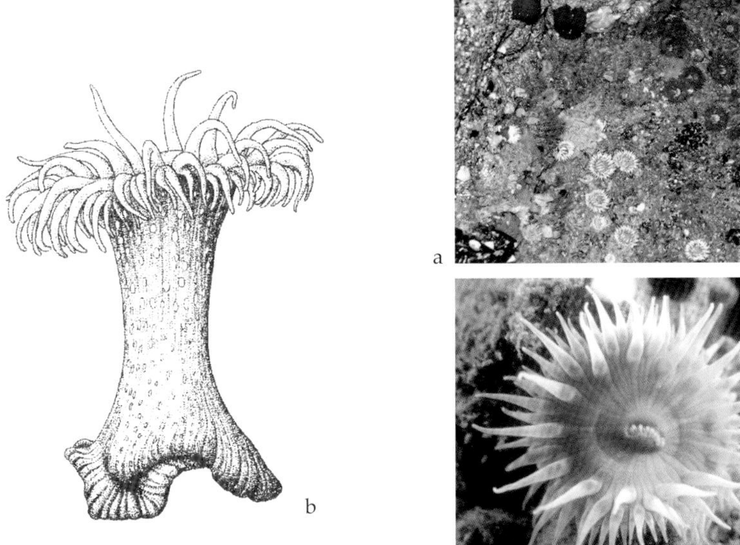

L.11 *Sagartia elegans*: **a** four colour varieties in a single pool. (Photo: J.S. Ryland); **b** to show column; **c** variety *venusta*. (Photo: D.N. Huxtable/J.S. Ryland)

in small groups in pools from MTL down, all coasts of Britain and Ireland *Sagartia elegans*

Five colour varieties of this small anemone are recognised: var. *nivea*: disc and tentacles uniformly white; var. *rosea*: disc colour variable, tentacles rose-pink to magenta; var. *venusta*: disc orange, tentacles white; var. *aurantiaca*: disc colour variable, tentacles dull orange; and var. *miniata*: disc and tentacles patterned brown and orange, in varying shades.

Sagartia ornata is an even smaller species, the column less than 15 mm diameter, with inconspicuous suckers. Tentacles have pale banding, in four to five cycles; column translucent green to brown, oral disc with white spots. Lower shore in crevices; probably present on all coasts of Britain and Ireland. It was formerly confused with *Sagartia troglodytes*, a larger species (up to 12 cm high), with around 200 tentacles, arranged in cycles of six; colour variable, disc and tentacles with dark patterning, distinct at base of each tentacle. Intertidal, to MTL, and shallow sublittoral; all coasts.

11. Column tall, slender, flaring to disc, up to 12 cm high; around 100 long tentacles, with span of 15 cm, which usually do not fully retract to disc. Cinclides conspicuous when column is fully extended, but acontia are not readily emitted. Dull brown with white lines on disc (L.12). Lower shore; rare, south-west coasts from Dorset to Bristol Channel, also Channel Isles *Aiptasia couchii*

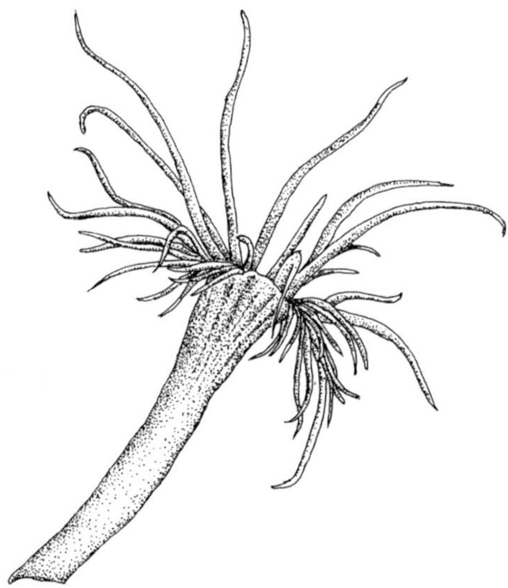

L.12 *Aiptasia couchii.*

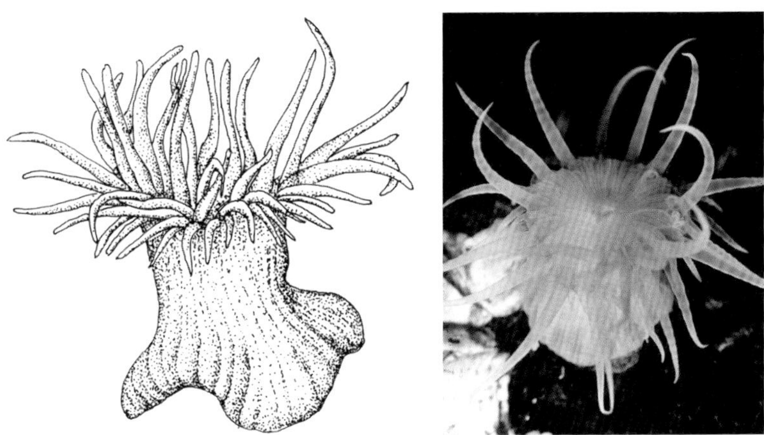

L.13 *Actinothoe sphyrodeta*: **a** to show column; **b** view of disc and tentacles. (Photo: D.N. Huxtable/J.S. Ryland)

– Column short and broad, up to 2 cm high, 15 mm diameter, with up to 120 irregularly arranged tentacles (L.13). Cinclides apparent as dark spots, acontia freely emitted. Disc white or orange, with white tentacles: similar to *Sagartia elegans* var. *venusta*, but distinguished in lacking suckers. Low-shore pools, from east Channel to Shetland *Actinothoe sphyrodeta*

Key M Echinoderms

Few species of echinoderm – starfish, brittle stars and sea urchins – occur commonly on rocky intertidal seashores. The most familiar starfish, *Asterias rubens*, is common to abundant off all north-west European coasts and at times huge numbers may accumulate briefly between the tides; sporadically, the Edible Sea Urchin *Echinus esculentus* (M.1) may also occur in numbers amongst dense kelp beds at ELWS.

Neither of these species can be considered part of the rock-pool fauna, but of eight species occurring frequently on intertidal rocky shores, two qualify as permanent residents. All echinoderm species recorded from the north-west European shelf are described and illustrated by Southward & Campbell (2006). The following diagnostic descriptions apply **only** to the eight species keyed here. The keys provided by Southward & Campbell (2006) should be consulted for the identification of any other species encountered.

Echinoderms characteristically display a pentamerous (five-rayed) symmetry, although it may be overlain by a secondary bilateral symmetry, as in the heart urchins. This is apparent from the tube feet, employed in feeding, walking or securing the animal to its substratum.

tube feet
retractile, flexible, membranous tubes, often with terminal suckers, extended through pores in some skeletal plates; they occur in rows beneath the arms of starfish and brittle stars, and in five defined ambulacral rays in urchins

1. Brittle stars (Ophiuroidea): animals with five slender, flexible arms, distinct from, and much longer than, the rounded central disc. Identification of ophiuroids depends upon examination of the calcareous units – plates and spines – comprising the segments of the arm, and those of the often complex jaw apparatus

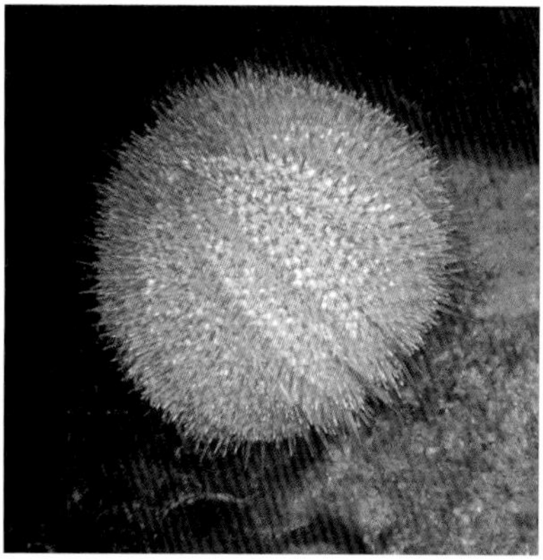

M.1 The Edible Sea Urchin *Echinus esculentus*. (Photo: P.E.J. Dyrynda)

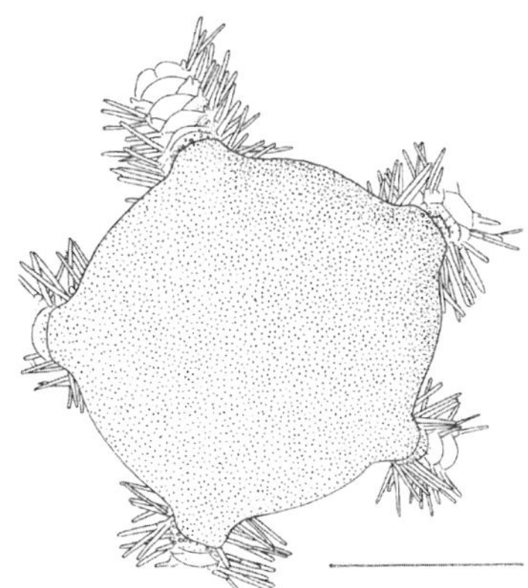

M.2 *Ophiocomina nigra*: disc and arm bases in dorsal view, to show uniformly granular surface. Scale bar: 10 mm.

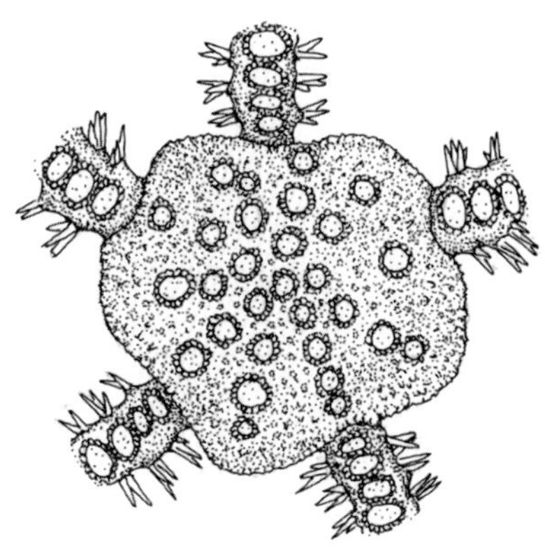

M.3 *Ophiopholis aculeata*: disc and arm bases in dorsal view, showing pattern of larger plates bordered by small plates, on a granular ground. Scale bar: 5 mm.

surrounding the ventral mouth. However, four species found in the rocky intertidal may be recognised by sizes and patterns of the plates forming the upper (dorsal) surface of the disc **2**

– Not brittle stars **5**

2. Dorsal surface of disc completely covered with small uniform granules, no larger scales or plates. Disc up to 25 mm diameter, arms to 12 cm, each arm joint with five to seven smooth, slender spines on each side. Colour variable, entirely black to brown, yellowish, orange or white, often patterned (M.2). Essentially sublittoral, but may occur at ELWS, under stones, in deep pools. West Channel to Shetland, north-east England; all coasts of Ireland *Ophiocomina nigra*

– Dorsal surface of disc not uniformly granular, with some larger scales or plates **3**

3. Dorsal surface of disc with a pair of larger plates – the radial shields – close to the base of each arm **4**

– Without radial shields. Disc surface covered with fine granules and small spines, and with larger, rounded plates spaced apart. Disc up to 2 cm diameter, arms to 8 cm, each arm joint with six or seven short, blunt spines on each side (M.3). Lower shore and sublittoral, most frequent on northern and western coasts, absent from English Channel and North Sea coasts *Ophiopholis aculeata*

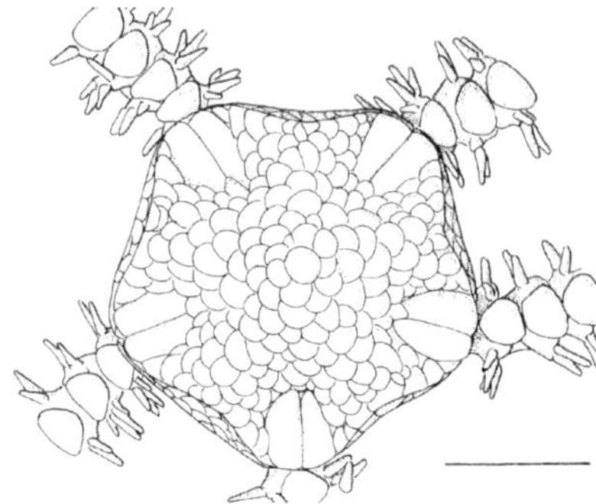

M.4 *Amphipholis squamata*: disc and arm bases in dorsal view, to show small radial shields in contiguous pairs. Scale bar: 1 mm.

4. Disc diameter up to 5 mm, radial shields small (less than one-quarter disc diameter), each pair close together. Arms up to 2 cm, each arm joint with two or three short, blunt spines. Pale brown, grey to orange, or reddish. A tiny species, often abundant amongst *Corallina officinalis* and in red algal turf (M.4); also sublittoral. All coasts in the region *Amphipholis squamata*

– Disc diameter up to 2 cm, radial shields large (more than one-third disc diameter), each pair spaced apart. Arms up to 10 cm, each arm joint with seven or eight serrated spines on each side. Colour variable, red, green, yellow, violet, the arms often distinctly colour banded (M.5). Lower shore and sublittoral, all coasts in the region *Ophiothrix fragilis*

5. Starfish (Asteroidea): animals with five or more arms, short and broad based, or long and tapered, but always contiguous at the base and not distinctly separate from

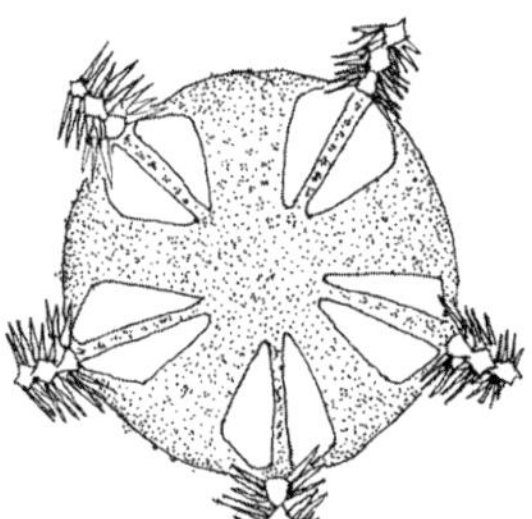

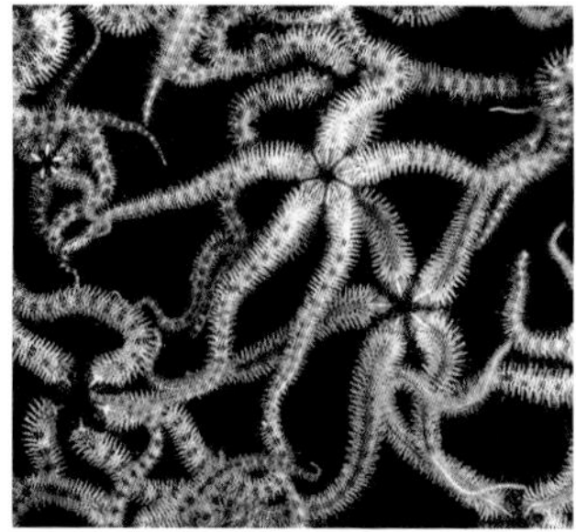

M.5 *Ophiothrix fragilis*: **a** disc and arm bases in dorsal view, to show large radial shields, each pair spaced apart (scale bar: 5 mm); **b** individuals in ventral (oral) view.

M.6 The Common Cushion Star *Asterina gibbosa*. (Photo: J.S. Ryland)

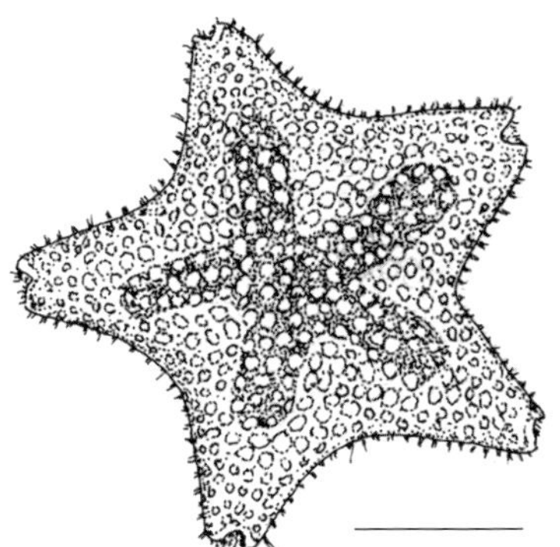

M.7 The characteristically patterned *Asterina phylactica*. Scale bar: 5 mm.

the central disc. Most species are strictly sublittoral in distribution, apart from *Asterias rubens* and *Astropecten irregularis*, which can be found around ELWS on sheltered sandy coasts, but two small cushion stars occur frequently in rocky intertidal habitats:

– Arms very short, the entire animal having a broadly stellate (star-shaped) outline, up to 6 cm diameter; light brown, yellow to orange, to grey green, and either unicolorous or mottled (M.6). In pools, under stones or overhangs, lower shore and sublittoral. West coasts of Britain, from west Channel to Shetland, north and west Irish coasts *Asterina gibbosa*

– Body similarly proportioned, but maximum diameter only 15 mm, and dull green in colour with a dark brown central star pattern (M.7), juveniles (<4 mm) lack the star and are indistinguishable from *A. gibbosa*. On exposed rocky shores, usually in deep pools above MTL, often in *Corallina* turf. Presently known from a few shores on south-west coasts of England, Wales and Ireland, and from Strangford Lough (Northern Ireland) and the Isle of Man *Asterina phylactica*

test
the rigid skeletal structure supporting the body wall in echinoids, consisting of calcareous plates, in rows from the upper (aboral) to the lower (oral) surface

– Sea urchins (Echinoidea): animals with globular or ovoid, spiny tests, slightly flattened in some species. Five-rayed symmetry expressed in radiating rows of spines, alternating with rows of tube feet. As noted above, *Echinus esculentus* may occur amongst kelp close to ELWS on some moderately exposed shores, but two smaller species may be found in pools up to MTL, or higher on exposed coasts:

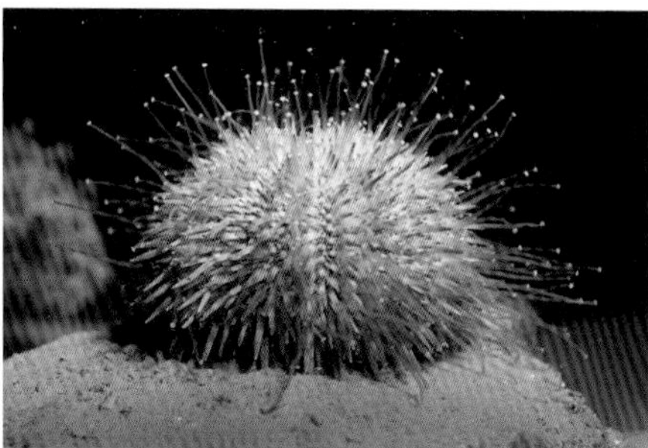

M.8 The Green Sea Urchin *Psammechinus miliaris*, with tube feet extended. (Photo: P.E.J. Dyrynda)

M.9 *Paracentrotus lividus* and *Actinia equina*, in an exposed-shore, coralline rock pool.

– Test a flattened ovoid, rather than globular, up to 5 cm diameter; green, with violet tips to the spines (M.8) (*E. esculentus* red to pink, with white, pink or violet spines). The ventral mouth is bordered by a membrane covered with thick plates (juvenile *E. esculentus* with scattered thin plates). In pools, often beneath stones, and sublittoral, all coasts within the region *Psammechinus miliaris*

– Test a flattened ovoid, up to 7 cm long; dull green, with deep violet to dark olive spines, often appearing black (M.9). Intertidal, typically in coralline rock pools on exposed coasts, and shallow sublittoral, in pits and crevices, and in cavities bored into soft rocks. South and west coasts of Ireland, rare Devon and Cornwall *Paracentrotus lividus*

Key N Rock-pool fishes

The most extensive intertidal rocky habitats support or shelter an interesting variety of fish species, some individuals of which may be surprisingly large. Mostly, they will be bottom-dwelling species associated with rocky grounds or mixed muddy bottoms, including a number largely limited to intertidal or shallow coastal habitats, but others will be essentially pelagic in habit, associated with kelps and other large brown seaweeds. Some may be resident through much of the year while others spend a brief time inshore during spring and summer. Dogfish *Scyliorhinus canicula* often attach their distinctive egg cases – the familiar Mermaid's purse – to large brown seaweeds, and occasionally an individual may become temporarily stranded in a low-shore pool through the emersion period. Lumpsuckers *Cyclopterus lumpus* deposit their eggs in sheets in shallow rocky habitats; the brightly coloured male attends the eggs until they hatch, and can be found in deep low-shore pools in early spring. Intimidatingly large Conger Eels *Conger conger* often take up residence in deep crevices in pools as far upshore as MTL (see Fig. 2.4); several small species of wrasse establish territories during the breeding season along rock ridges at ELWS (see Fig. 2.4) and, in late summer, pools from MTL down may provide refuge for the shoals of juvenile Herring *Clupea harengus* and Sprats *Sprattus sprattus*, which are collectively termed 'whitebait'.

The number of fish species recorded for a moderately exposed, weedy rocky shore is likely to reflect the persistence and diligence of the recorder, through the seasons and in relation to the weather. It will encompass a broader or narrower selection of the local inshore fish fauna, together with seasonal migrants and occasional vagrants. However, a small number of species may be regarded as characteristic of rocky-shore habitats, including rock pools, with representatives of families especially adapted for life in the intertidal. There are numerous guidebooks devoted to the identification of marine fish, and a mine of images and descriptions is available online. The following key will assist identification of the most common rocky shore fishes known for the region. **However**, if a specimen does not accord with any of the key characters given here, it should be checked against a more comprehensive account.

1. Pipefishes (Syngnathidae): long, thin and rather stiff-bodied fish, with a single, short dorsal fin; with a small, round mouth at the end of a slender snout (N.1–N.3). Beneath stones amongst weed, generally slow moving **2**

– Not pipefishes **4**

2. With a distinct caudal (tail) fin and well-developed pectoral fins. Snout large, much longer than rest of head; body encased in conspicuous bony rings. Up to 46 cm long, brown, with dark banding (N.1). ELWS, amongst seaweed and seagrass; all coasts within the region, often common Greater Pipefish *Syngnathus acus*

   Two similar, but less common, species are *S. rostellatus*, in which the snout is about the same length as the rest of the head, and *S. typhle*, in which the snout is laterally flattened, and at its tip as deep as the rear part of the head.

– Body tapered posteriorly, without a caudal fin **3**

N.1a, b The Greater Pipefish *Syngnathus acus*. Scale bar: 20 mm. (**a** photo: P.E.J. Dyrynda)

N.2 The Worm Pipefish *Nerophis lumbriciformis*. Scale bar: 20 mm.

N.3. The Straight-nosed Pipefish *Nerophis ophidion*. Scale bar: 20 mm.

3. Dorsal fin with 24 to 28 fin rays. Snout short and upturned. Up to 15 cm long, dark brown, with white streaks and patches on head and just posterior to it (N.2). Usually under stones, amongst silt, lower shores. All coasts, common
   Worm Pipefish *Nerophis lumbriciformis*

– Dorsal fin with 34–40 fin rays. Snout straight, about as long as rest of head. Up to 30 cm long, green, paler ventrally, sometimes pale bluish around head (N.3). Lower shore, amongst seaweed and seagrass. All coasts, patchily distributed
   Straight-nosed Pipefish *Nerophis ophidion*

4. Ventral surface of fish with a conspicuous adhesive disc (N.4a, N.6a, N.11a), formed from the conjoined pelvic fins **5**

– Without a ventral adhesive disc **12**

5. Adhesive disc sucker-like, with entire, undivided rim (N.4a); single, long dorsal and anal fins. Skin without scales, slippery or slimy to touch. Dull brown, to 6 cm long. ELWS and shallow sublittoral, associated with large seaweeds, often found in hollow holdfast of the kelp *Saccorhiza polyschides*. All coasts within the region Montagu's Sea Snail *Liparis montagui*

– Adhesive disc simple (N.6a), or complex and divided, with clefts and folds (N.11a), but rim discontinuous, not sucker-like. With one or two short dorsal fins, and one short anal fin. Scales present or absent **6**

6. Bull-headed, with prominent eyes. Scales present; with two dorsal fins, anterior one the shorter **7**

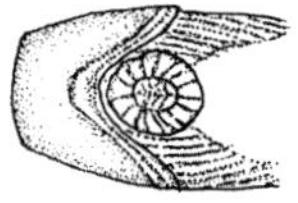

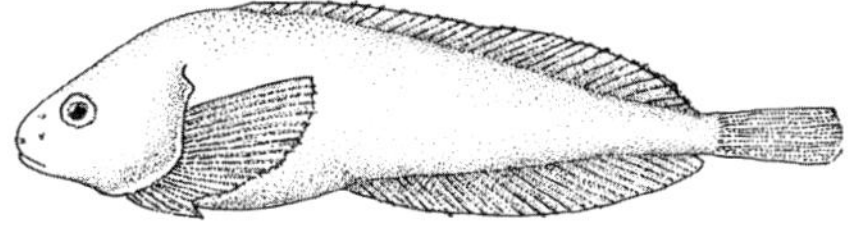

N.4 Montagu's Sea Snail *Liparis montagui*: **a** the ventral adhesive disc; **b** in lateral view.

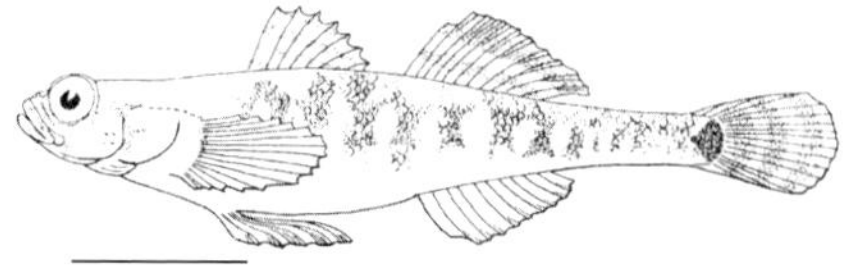

N.5 The Two-spot Goby *Gobiusculus flavescens*. Scale bar: 10 mm.

– Head flattened, with tapered snout. No scales; only one dorsal fin **10**

7. First dorsal fin with seven or eight spine rays. A conspicuous black spot at the base of the tail, on each side. Eyes laterally situated. Reddish brown to green, palest ventrally, with light banding along flanks and on dorsal fins. To 6 mm long (N.5). Intertidal and shallow sublittoral, amongst algae and in weedy pools. All coasts within the region Two-spot Goby *Gobiusculus flavescens*

– First dorsal fin with six or seven spine rays. No black spot at tail base. Eyes dorso-laterally situated, appearing to be on top of head **8**

8. First (i.e. upper) four or five rays of the pectoral fins filamentous rather than spiny, and not attached to the fin membrane (N.6a). Dark brown to almost black, the first dorsal fin with a yellow to orange marginal band. To 12 cm long. Intertidal and shallow sublittoral, may occur in pools well above MTL. All coasts, common. Rock Goby *Gobius paganellus*

This goby is ubiquitous in rocky intertidal habitats throughout the region. The equally large Leopard-spotted Goby *Thorogobius ephippiatus* also occurs on all coasts of Britain and Ireland, though scarce on south-east shores of the North Sea, but is often overlooked because of its retiring habit, occupying deep crevices in vertical

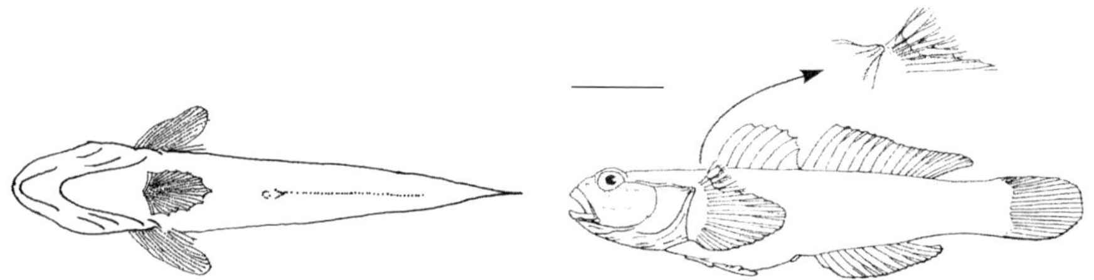

N.6 The Rock Goby *Gobius paganellus*: **a** diagram to show ventral pelvic sucker; **b** in lateral view; detail shows first fin rays of the pectoral fin. Scale bar: 10 mm.

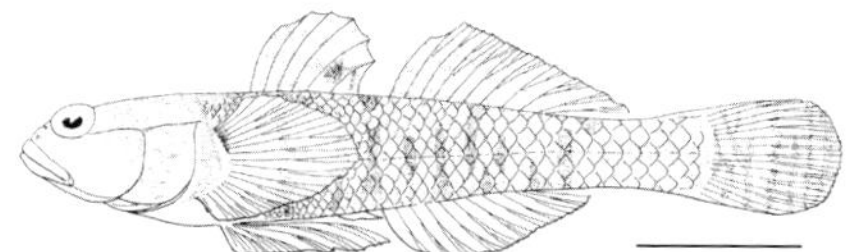

N.7 The Common Goby *Pomatoschistus microps*. Scale bar: 10 mm.

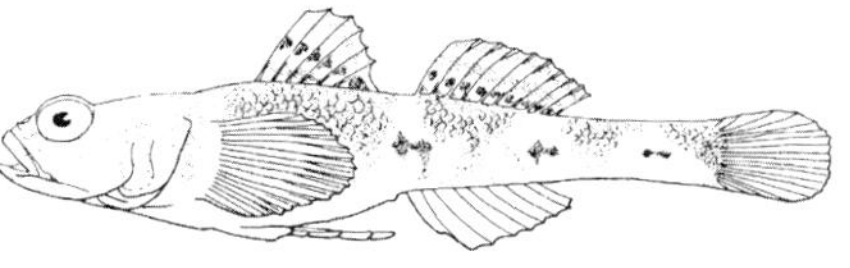

N.8 The Painted Goby *Pomatoschistus pictus*. Scale bar: 10 mm.

rock faces. Once seen it is easily recognised by its light brown coloration and large orange or black spots.

– All rays of the pectoral fins spiny, not filamentous **9**

9. Each longitudinal row has 40–50 scales, from behind head to tail base. Grey to buff, with a few vertical dark bars on the flank; male with a dark blotch at the lower rear edge of the first dorsal fin, but no other marking on either dorsal fin. Up to 7 cm long (N.7). Intertidal, in all coastal habitats; on rocky shores occurring in pools well above MTL. All coasts in the region
Common Goby *Pomatoschistus microps*

– Each longitudinal row has 35–40 scales. Brown, with four darker patches dorsally, from first dorsal fin to caudal fin, and four diffuse spots along flank; both dorsal fins with rows of black spots and longitudinal red bands. Up to 5.5 cm long (N.8). On rocky grounds, occasionally at ELWS. All coasts in the region
Painted Goby *Pomatoschistus pictus*

10. Dorsal and anal fins short and similarly sized, well separated from the caudal fin, and each with four to seven spiny rays, the first anal fin ray more or less below the first dorsal ray. Female with small dark spot at origin of dorsal and anal fins, male with dark spot on each. Greenish brown, up to 5 cm (N.9). Lower shore, amongst weed, often in holdfast of *Saccorhiza polyschides*. South and west coasts of Britain
Small-headed Clingfish *Apletodon dentatus*

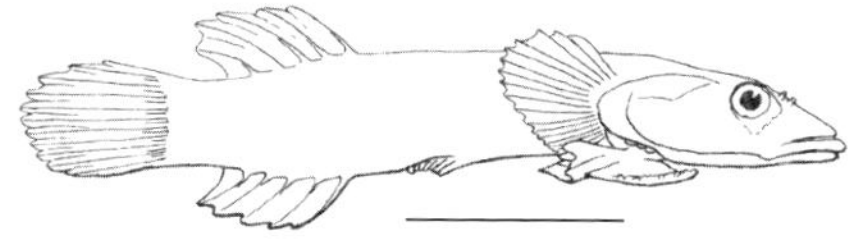

N.9 The Small-headed Clingfish *Apletodon dentatus*. Scale bar: 10 mm.

The Two-spotted Clingfish *Diplecogaster bimaculata* is similar, but reddish in colour; it lacks black spots on the dorsal and anal fins, but males have a pair of large purple and yellow spots ventro-laterally, close to the pectoral fins. All coasts, rarely lower shore.

– Dorsal fin with 15–20 spiny rays, anal fin with 9–12 **11**

11. Dorsal and anal fins continuous with caudal fin. Head broad and flattened, with a pair of conspicuous tentacles at the front. Light reddish brown, with large darker red spots and broader stripes extending anteriorly and postero-ventrally from each eye; a pair of conspicuous red-ringed, blue spots dorsally, behind head. Up to 7.5 cm long (N.10). Low shore, amongst weed on rocky coasts. South and west coasts from Dorset to Shetland, all coasts of Ireland
Cornish Clingfish *Lepadogaster purpurea*

– Dorsal and anal fins not continuous with caudal fin. Head broad and flattened, with inconspicuous tentacles. Colour very variable: males often reddish brown with darker spots and blotches; females greenish with spots and streaks. Up to 10 cm long (N.11). ELWS and below, amongst algae on rocky shores. South and west coasts, from Devon to Scotland, west Ireland
Connemara Clingfish *Lepadogaster candolii*

barbel
a slender sensory tentacle, present in many bottom-dwelling fish

12. Fin rays soft, **not** spiny. Two dorsal fins and one anal fin present. With barbels at tips of either or both upper and lower jaws, and close to nostrils **13**

– Both soft and spiny fin rays present. Only one dorsal fin apparent, some isolated spines may be present anterior to it **14**

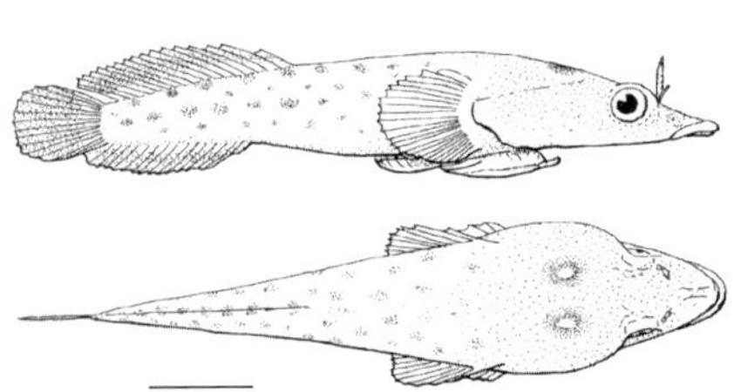

N.10 The Cornish Clingfish *Lepadogaster purpurea*. Scale bar: 10 mm.

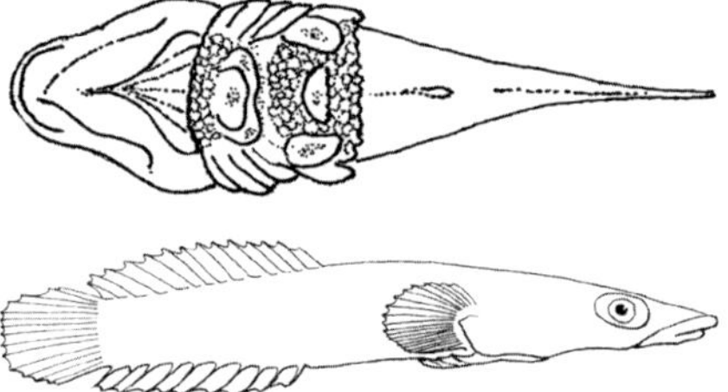

N.11 The Connemara Clingfish *Lepadogaster candolii*: **a** diagram to show complex ventral sucker; **b** in lateral view. Scale bar: 10 mm.

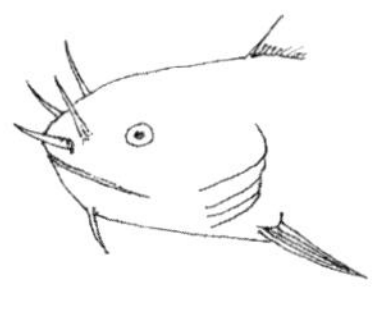

N.12 The Five-bearded Rockling *Ciliata mustela*: **a** photo: J.R. Ellis/CEFAS; **b** diagram of head, showing five barbels.

13. Head with one barbel on lower jaw, two on upper jaw and one on each nostril. Dorsal fin beginning with a single long ray and a series of short individual rays. Dark brown above, white below. Up to 30 cm long (N.12). Intertidal, below stones, amongst weed; common on all coasts in the region

    Five-bearded Rockling *Ciliata mustela*

– Head with three barbels only, one on lower jaw, one by each nostril. Dark brown above, paler below. Up to 25 cm long (N.13). Intertidal, below stones, amongst weed, common on all coasts

    Shore Rockling *Gaidropsarus mediterraneus*

    The Three-bearded Rockling *G. vulgaris* is similar, but larger (up to 60 cm) and light pinkish brown with darker brown spots. It occurs throughout the region but only rarely on shore.

14. First dorsal fin represented by a series of separated spines; the single prominent fin is the second dorsal fin:

– Fifteen short spines in front of the second dorsal fin; body long and slender, markedly tapered towards the caudal fin. Olive brown above, lighter below, sometimes with dark markings on flanks. Up to 20 cm long (N.14). Inshore, amongst seaweed and seagrass. All coasts, less common in south-east North Sea

    Fifteen-spined Stickleback *Spinachia spinachia*

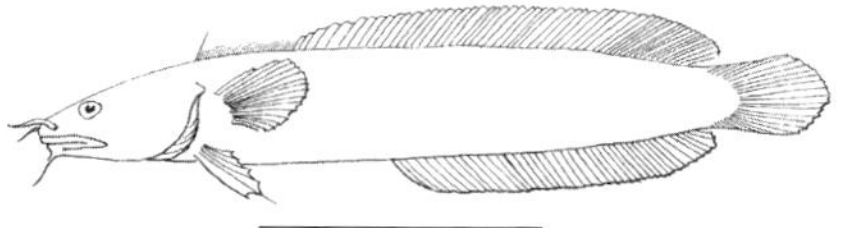

N.13 The Shore Rockling *Gaidropsarus mediterraneus*. Scale bar: 50 mm.

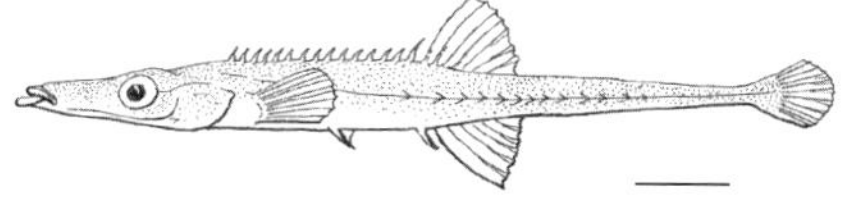

N.14 Fifteen-spined Stickleback *Spinachia spinachia*. Scale bar: 10 mm.

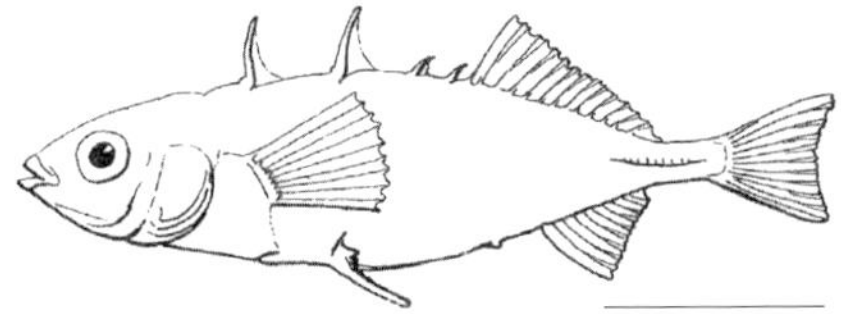
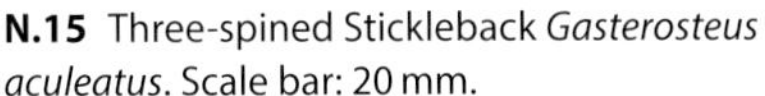

N.15 Three-spined Stickleback *Gasterosteus aculeatus*. Scale bar: 20 mm.

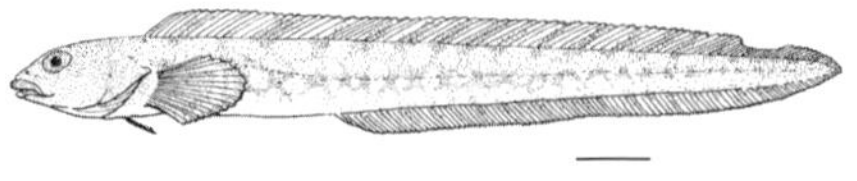

N.16 The Viviparous Blenny *Zoarces viviparus*. Scale bar: 10 mm.

– Two long spines and one much shorter spine (or two), well spaced, in front of the second dorsal fin. Body short and plump. Greenish brown above, with darker lateral bands, silvery white below (red in breeding males). Up to 10 cm long (commonly smaller) (N.15). Mainly fresh water, but may occur in brackish, weedy pools on all coasts. Three-spined Stickleback *Gasterosteus aculeatus*

– Only a single dorsal fin present **15**

15. Dorsal and anal fins both very long, continuous with the caudal fin; a series of short, spiny rays at posterior end of dorsal fin. Body long, slender, rather eel-like. Brown above, paler below, with darker patches dorsally, extending on to dorsal fin and along the flanks. Up to 35 cm long (N.16). Intertidal, moving offshore during summer; a northern species, ranging from Irish Sea to Scotland and to North Sea coasts
Viviparous Blenny *Zoarces viviparus*

– Dorsal and anal fins long or short, but neither continuous with the caudal fin **16**

16. Dorsal fin long, extending almost to base of caudal fin, with short spiny fin rays only. Pelvic fins minute. Body long, slender, eel-like. Yellowish brown, with 9–15 white-ringed black spots along base of dorsal fin. Up to 25 cm (N.17). Intertidal and shallow sublittoral, under stones, amongst algae. All coasts Butterfish *Pholis gunnellus*

– Dorsal fin not uniform, but with two distinct regions consisting of spiny or soft rays **17**

17. Anal fin with three (rarely six) spiny rays and 5–14 soft rays. Dorsal fin with continuous edge, the anterior two-thirds with sharply pointed spiny rays, the posterior with soft rays. Up to 15 cm long. Very colourful: shades of blue, yellow or green, with blue and orange stripes on

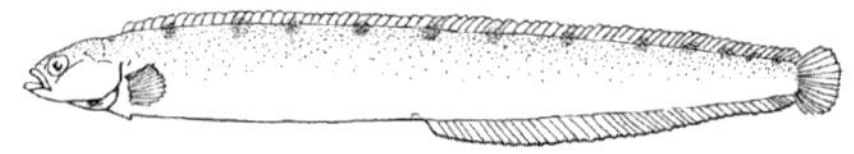

N.17 The Butterfish *Pholis gunnellus*. Scale bar: 50 mm.

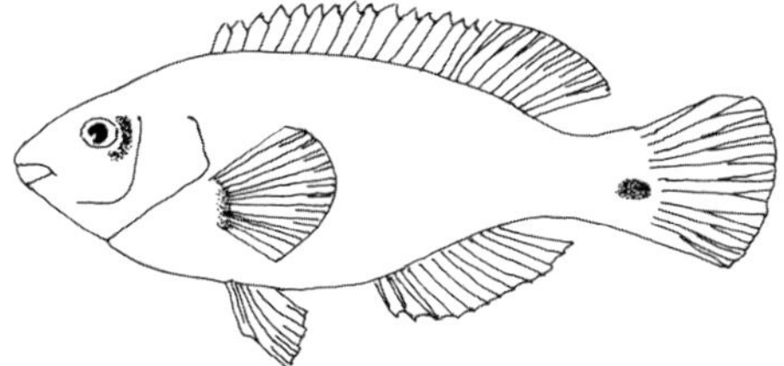

N.18 The Corkwing *Symphodus melops*.

head; a large black spot close to the base of the caudal fin (N.18). In deep, weedy pools close to ELWS, also shallow sublittoral Corkwing *Symphodus melops*

The Goldsinny *Ctenolabrus rupestris* is similar but less colourful, being reddish brown with a dark spot at the anterior of the dorsal fin and another dorsally close to the caudal fin. Several other species of wrasse occur in shallow sublittoral habitats, especially off south-west coasts. However, colours vary with size and age, sex and season, and are not the best guide to identification; morphological features, including counts of fin rays and scales should be employed. (See, e.g., Hayward & Ryland 2017.)

– Anal fin with two spiny rays and 15–26 soft rays. Dorsal fin divided into two more or less equal parts by a distinct notch. The Common Blenny or Shanny *Lipophrys pholis* occurs abundantly in intertidal rocky shore habitats throughout the region. Two rarer species may be found, mostly on south-west coasts, in large, low-shore pools; they are readily distinguished by the presence of fringed flaps or tentacles on the head:

– Head smooth, lacking tentacles. Colour very variable: green, brown, reddish or black, with lighter and darker blotches and streaks. Up to 16 cm (N.19). Intertidal, ubiquitous on rocky coastlines, occurring in the smallest pools well above MTL, and often found sheltering in damp crevices during emersion periods
Common Blenny or Shanny *Lipophrys pholis*

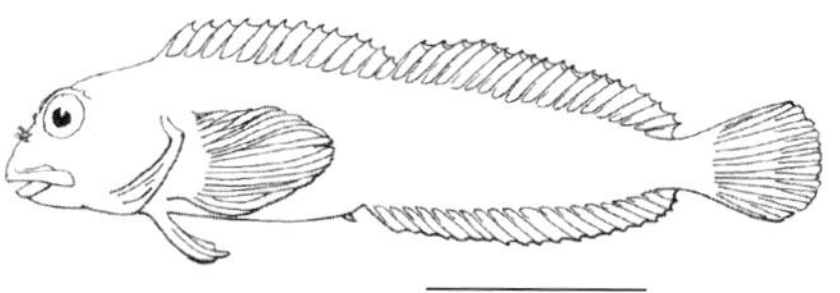

N.19 The Common Blenny or Shanny *Lipophrys pholis*: **a** photo: J.R. Ellis/CEFAS; **b** diagram showing detail of fin rays. Scale bar: 50 mm.

– Head with an erectile, fringed flap anteriorly, with a series of short, filiform tentacles behind it. Shades of brown, with darker bands dorsally and blue-white spots along flank. Up to 8 cm (N.20). Intertidal, in pools around MTL. South-west Britain (Dorset to Pembrokeshire), south and west Ireland
Montagu's Blenny *Coryphoblennius galerita*

– Head with a branching, filiform tentacle above each eye. Brown, with darker, broad, vertical bands. Up to 30 cm long (N.21). At ELWS on weedy shores. South and west coasts of Britain, to Shetland, sparse on east coasts; all Irish coasts Tompot Blenny *Parablennius gattorugine* (see cover photograph)

N.20 Montagu's Blenny *Coryphoblennius galerita*. Scale bar: 50 mm.

N.21 Tompot Blenny *Parablennius gatturogine*. Scale bar: 50 mm.

6 Investigating rock pools

The ecology of rocky seashore communities has attracted considerable research interest through many decades, much of which has been gathered into substantial texts, such as *The Biology of Rocky Shores* (Little *et al.* 2009), or as dedicated chapters in broader marine biology reviews, such as *Marine Ecology: Processes, Systems, and Impacts* (Kaiser *et al.* 2020). Research into the interplay of physical environmental factors and biological processes in determining the structure and composition of rocky shore communities continues, and is likely to increase as the consequences of climate change begin to impact coastal habitats. Rock-pool habitats are a significant part of the intertidal ecosystem on most rocky coasts but they have rarely been the main focus of research. For the most part, rock pools have been considered in the broader context of the ecology and distribution of the dominant intertidal fauna and flora, the zoned macroalgae, the dominant grazers (winkles, limpets and topshells) and the most significant predators (dog whelks and shore crabs). There have been few studies of the physical environmental characteristics of rock pools comparable, for example, to those of Daniel & Boyden (1975), few studies of rock-pool communities comparable to those of Björk *et al.* (2004), Davenport *et al.* (1997), Davenport *et al.* (2004), Hull (1997, 1999a, 1999b) or McAllen (1999, 2001), and only few population studies of the small gastropods so common amongst coralline/red algal pool habitats (e.g. Wigham 1975; Fretter & Manly 1977; Southgate 1982). There is considerable potential for new and original contributions to these particular fields, and much that could be achieved with the simplest techniques.

Each rock pool is a unique microcosm, defined by its particular physical and biological characteristics, and field work should be conducted with this constraint as a priority. It is important, obviously, to be certain that the shore to be visited is not a protected site, and damaging fragile communities, such as reefs of *Sabellaria* (honeycomb worms), through careless trampling should be avoided. Modifying the environment of a pool, particularly isolated pools on the middle and upper shore, may be damaging. Some techniques previously employed, such as baling out or de-vegetating pools, are best forgotten. For comparative purposes, the tidal level of each pool studied should be recorded and the degree of wave exposure to which it is subject estimated. Tidal level may be approximated by using local tide tables, available

Fig. 6.1 Dense bands of *Fucus* are seen above MTL on sheltered shores: *F. spiralis* is succeeded, below, by *F. vesiculosus*.

in most coastal towns, which provide times and heights of daily tides, and more accurately through the Admiralty Tide Tables (available online) which supply tidal predictions and curves for standard ports around the British Isles. In practice, it is acceptable to record simply whether the pool studied is situated on the lower, middle or upper shore, by reference to the local tide table. There is no simple way to measure the degree of wave exposure a shore, or part of a shore, may experience. The biological exposure scale devised by Ballantyne (1961) grades shores on a scale of 1 (very exposed) to 8 (very sheltered) on the basis of the varying abundance of indicator species along a down-shore transect. It is useful in demonstrating that the composition of communities and the relative abundance of species change across horizontal gradients, but provides no meaningful measure of wave exposure. However, it does provide a useful shorthand: shores dominated by zoned algal communities MTL (Fig. 6.1) are at the sheltered end of the gradient, those dominated by barnacles, mussels and limpets are at the exposed end (Fig. 6.2), whilst 'moderately exposed/semi-sheltered' shores are characterised by mosaic communities of fucoid algae, barnacles and limpets (Fig. 6.3).

Fig. 6.2 An upper-shore pool on a wave-exposed promontory dominated by barnacles and limpets.

Fig. 6.3 'Mosaic' communities on a semi-exposed shore, below MTL, with patches of *Fucus serratus*, red algae, barnacles and limpets.

The surface area, depth and volume of each pool studied should be recorded. Surface area may be computed from digital images taken above a rigid grid. Where depth appears to vary little, a maximum value is probably sufficient, but where a steep depth gradient is evident it is more useful to calculate an average from a series of measures. Depth measurements, together with length and breadth, can provide rough estimates of pool volume. (The technique of measuring the volume of water baled out from a pool is likely to cause environmental stress and is inadvisable.) Volume can be estimated from a maximum or average depth estimate (*d*), together with measurements of length (*l*: the maximum distance across the pool) and width (*w*: the distance perpendicular to the length at its midpoint). These three values are then used to calculate the volume of a solid shape that approximates to the form of the pool (White *et al.* 2015). Thus, the volume of a circular pool of constant depth may be estimated by the equation for a cylinder, $\pi r^2 l$, where $r = 0.5\ w$ and $l = d$; for an oval pool shallowing at each end the equation for a semi-ellipsoid ($0.083\pi lwd$) provides an approximate volume. The significance of the volume of water held to the ecology of a pool is not always clear; depth seems to be the most important factor determining the structure of rock-pool algal communities (Martins *et al.* 2007), and fish assemblages (White *et al.* 2015), although both volume and depth vary in effect in relation to tidal level.

Temperature, salinity and pH fluctuate during tidal emersion in relation to pool size and depth, and to tidal level. Ideally, the physical environmental profile of a pool investigated should include measurements of each of these parameters through diurnal and nocturnal emersion cycles, at both high and low extremes of the local tidal cycle, and at seasonal intervals. This is a less onerous task than might appear; portable temperature/salinity meters, pH meters and combined T/S/pH meters are standard equipment for most field centres and survey groups, and are also relatively inexpensive to purchase.

Large seaweed species and most of the mobile macrofauna found in or adjacent to pools may be identified in the field using a standard seashore guide. Depending upon the topography of the pool, it may be possible to record numbers of seasonally resident animals – in particular, fish and decapods – by regular sampling with a dip net at intervals through successive tidal cycles. Common Blennies *Lipophrys pholis* and Butterfish *Pholis gunnellus* may occupy the same pool throughout their summer sojourn in the intertidal, or may move between pools, and the hermit crab *Pagurus bernhardus* appears to lead a pretty nomadic existence. It is possible, and acceptable, to mark the shells of hermit crabs, topshells and winkles so that individuals may be recognised, and their distribution and movements on the shore, in and out of pools, may be recorded. A companion volume – *Snails on Rocky Sea Shores, Naturalists' Handbooks 30* (Crothers 2012) – provides excellent advice on techniques and methods for the study of epilithic gastropods. Hundreds of studies have been devoted to the ecology of limpets and winkles, from many localities on north-west European coasts, primarily because of their importance in contributing, through grazing, to the structure of rocky intertidal communities. However, in the case of the few species of small, epiphytic gastropods in which population biology and ecology have been studied, for example *Rissoa parva* (Wigham 1975) and *Barleeia unifasciata* (Southgate 1982), the published data derive from single programmes, at single sites, which have seldom been repeated.

The distribution and occurrence of probably a majority of the algal-associated invertebrates recorded from rock-pool habitats are incompletely known. Some species, while fluctuating in abundance seasonally, are constantly present, others are largely transitory, and in both cases presently known geographical distributions may be changing radically in a warming climate. Species records, ideally supported by clear digital images or preserved specimens, are usually welcomed by local biological recorders, and may often be submitted online; local museums and natural history groups are best approached for details regarding county recorders. Long-term monitoring of species and communities are also valuable in providing baselines for population studies, and may provide good project topics. The dearth of biological information on the small, algal-associated gastropods, as well as other epiphytic invertebrate species, needs to be resolved, through research across broader geographical and ecological ranges, and would be the most useful objective in studying rock-pool habitats. Turfs of *Corallina* and small red algae should be sampled quantitatively. Conventional advice is to use a small quadrat, scraping the sample into a plastic bag; in practice, two hands often prove insufficient for the task, with the result that a proportion of the quadrat sample is often lost. A moderately sized plastic bottle, with its base removed and the neck covered with a piece of fine mesh, makes both quadrat and collector (Fig. 6.4), and a flexible paint scraper requires just one hand.

Fig. 6.4 A makeshift 'bottle quadrat'.

$a = \pi r^2$ where $r = 0.5 d$

The sample may be recorded with reference to the diameter (*d*) of the bottle quadrat (thus, the area (*a*) sampled), total volume of material collected, or the wet weight of the weeds. Teasing out the samples in dishes of seawater in the laboratory will reveal a surprisingly rich fauna; left undisturbed, in cool conditions, most of the motile fauna of the sample will emerge, making for easier sorting, and sessile species such as tubeworms are then more readily found. Most of the turf fauna will require a microscope and reference to the specialist keys (Chapter 5) for accurate identification, although with practice some of the more distinctive gastropods may be readily distinguished with the use of a hand lens.

Samples should be sorted under a low-power stereomicroscope, using seekers, fine forceps and pipettes. Small crustaceans may be directly transferred to tubes of 70% methanol, but molluscs, polychaetes and cnidarians should be anaesthetised prior to identification and preservation; a 7.5% solution of magnesium chloride in seawater (i.e. 75 g per litre) is effective for molluscs and polychaetes, while menthol crystals scattered on the water surface will work for most cnidarians. In both cases, complete narcotisation may take a number of hours, during which the specimens should be kept cool. For permanent storage, soft-bodied specimens – polychaetes, cnidarians and unshelled heterobranchs – are best preserved in 4% seawater formalin, or following initial fixation in 4% formalin, in a 1% (by volume) aqueous solution of propylene phenoxetol. For everything else, 70% methanol is suitable. Magnesium chloride, menthol crystals and propylene phenoxetol are readily available from online suppliers, but methanol and formalin can only be used at established laboratories. Formalin is a 37% aqueous solution of the gas formaldehyde, a severe irritant that may result in damage to respiratory tissue if inhaled and cause painful irritation to skin and eyes; it should be handled under supervision and with extreme care. Colours usually fade swiftly following collection, and where they are important diagnostic characters, as in many sea slugs and polychaetes, a digital record should be made. It is good practice to retain all sorted samples, appropriately preserved and documented, as they constitute a permanent record as well as valuable reference sources.

Quantitative samples of coralline/red algal turfs can be analysed in many ways to provide insights into the ecology of the habitat. Total wet weight, and/or volume, of weed may be expressed per unit area sampled; wet weight and/or volume may calculated per unit area for each dominant algal species; the density of *Corallina* may be measured by reference to mean height of the fronds and frequency of branching (e.g. branches per cm). These data provide information on the complexity of the turf habitat. For the fauna, species richness may be expressed as total number of species per unit area, or per unit (weight or volume) of weed; measures of species diversity are calculated with reference to the relative proportions of each species in a sample, based upon numbers of species, number of individuals of each species and total number of all individuals. The most frequently used indices of diversity are based on the Simpson's Index, *D* (Table 6.1).

Measurements of shell length of all individuals for each gastropod species, plotted as size frequency histograms, provide insights into population structures (Fig. 6.5), and successive samples, at regular intervals, from contrasting pool types may yield interesting information on reproductive cycles, growth rates and longevity.

In most species of heterobranch, longevity is usually less than one year, while rock-pool populations are transient and often ephemeral. The most useful data to be collected for these animals are thus occurrence and density in relation to season and physical factors, especially temperature, salinity and pH. In some species, their appearance inshore perhaps coincides with spawning, and digital images of individuals associated with fresh spawn masses are also

Table 6.1 Calculating Simpson's Index, *D*.

From the data below, *D* is calculated as

$$D = \frac{\sum n(n-1)}{N(N-1)} = \frac{14832}{55932} = 0.265$$

where a value of 1 indicates no diversity, and 0 indicates infinite diversity. This is most usefully expressed as Simpson's Index of Diversity, $1 - D$, in which values range from 0 (no diversity) to 1 (infinite diversity) or as Simpson's Reciprocal Index, $1/D$, in which values increase with increasing diversity. Symbols: n = number of individuals of each species in the sample, N = total number of individuals in the sample and $\sum$ means sum the values for all species.

Species	n	$n - 1$	$n(n - 1)$
A	86	85	7,310
B	72	71	5,112
C	42	41	1,722
D	23	22	506
E	14	13	182
Total n (N)	237		
$N - 1$	236		
Total $n(n - 1)$			14,832

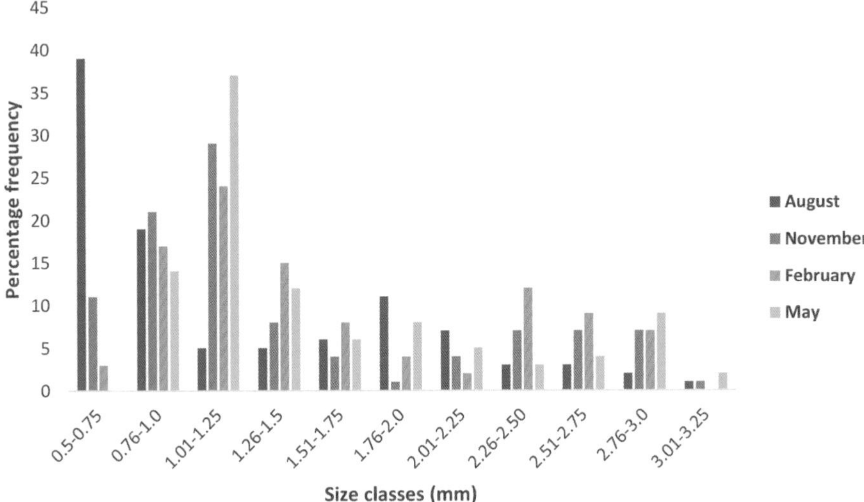

Fig. 6.5 Size frequency histograms showing population structure in the snail *Barleeia unifasciata* in four seasons. Peak summer recruitment is apparent in August, with a peak of older individuals indicating the previous year's recruitment. The two year classes are clear in November and February, but the older animals decrease in May. (Data from Southgate 1982)

useful. Samples of amphipod and isopod species may be sorted firstly by sex, and then by length, and the size and frequency of brooding females also recorded. For the two small echinoderms *Amphipholis squamata* and *Asterina phylactica*, size frequency distributions of disc diameter, and size and frequency of brooding individuals are also important population parameters; reproductive output is usually related to size of the brooding individual (Fig. 6.6).

Small polychaetes, particularly species of Syllidae, are the least well studied of algal-turf faunas, and the most difficult to process. All tend to be fragile and sensitive to the stress generated during sampling. Where individual species can be distinguished, the most useful information is often body length, and the number of body segments behind the head; however, such data are only of value for entire, undamaged specimens in which the terminal body segment can be recognised.

Ideally, all data resulting from field-based projects and monitoring programmes should be made available to other researchers, both amateur and professional. As noted above, there are many useful online discussion sites where data may be presented and exchanged. Occasionally, however, an investigation may yield new and original data of considerable

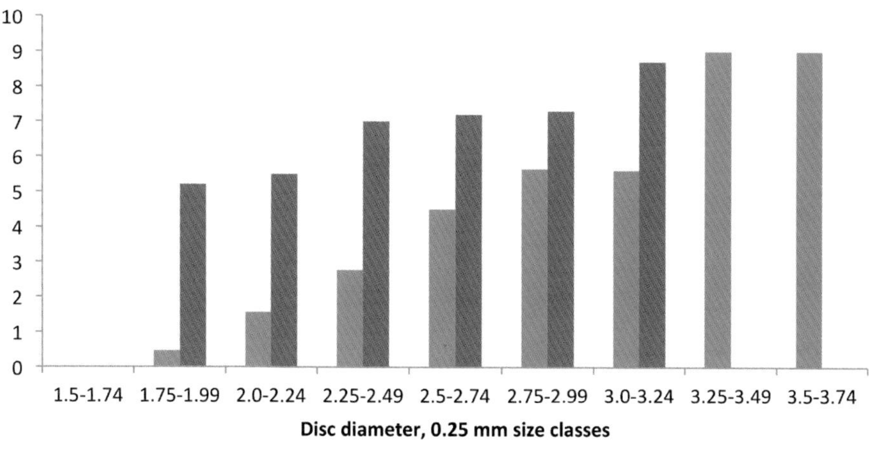

Fig. 6.6 Reproductive output in *Amphipholis squamata* in relation to disc diameter: number and size of embryos brooded increase with individual size. (Data from Emson & Whitfield 1989; embryo diameters were not recorded separately for the two largest size classes)

scientific significance. For example, the phenomenon of imposex, by which female dog whelks are effectively sterilised by the development of male sexual organs, was first observed by an undergraduate researcher (Blaber 1970), and ultimately revealed the effects and extent of pollution by tributyltin, derived from antifouling paints applied to boat hulls, in coastal waters (reviewed in Wigham & Graham 2018). Such cases may warrant publication in an appropriate journal, such as the *Journal of the Marine Biological Association of the United Kingdom*, as a short communication or a more detailed research paper. Guidelines for the preparation of manuscripts, and required formats, are available online for all scientific journals, and should be checked carefully before the submission of any manuscript.

7 References and further reading

Access to scientific literature is not always straightforward. All of the research journals listed here can be found online, but while the contents of each issue, together with abstracts of each paper, may be viewed, it is not always possible to view a paper of interest in its entirety. Some papers are published as open-access documents that may be downloaded, but for a majority a fee is usually demanded for this 'service'. Membership of a library that has an institutional subscription is then the only way to access papers and journals that do not provide open access. Similarly, while some more recent specialist identification keys are available online as open-access documents, others, such as the Linnean Society Synopses series, are only available in book form, and likely to be held only by university or field station libraries. Assistance with identification can be found online, with such sites as the Marine Life Information Network (MarLIN; www.marlin.ac.uk), but although most of these provide sound descriptions with diagnostic characteristics emphasised, clear illustrations and notes on distribution and occurrence, biology and ecology, they do not employ dichotomous keys. However, using the keys presented in this book may provide the necessary first steps in identifying problematic specimens using online sources.

Ashton, G.V., Brandt, A., Isaac, M.J., Makings, P., Moyse, J., Naylor, E., Smaldon. G. and Spicer, J.I. (2017) Crustaceans (phylum Arthropoda, subphylum Crustacea) In P.J. Hayward and J.S. Ryland (eds), *Handbook of the Marine Fauna of North-West Europe*, 2nd edn. Oxford: Oxford University Press.

Athersuch, J., Horne, D.J. and Whittaker, J.E. (1989) *Marine and Brackish Water Ostracods (Superfamilies Cypridacea and Cytheracea)*. Synopses of the British Fauna (New Series) 43. Leiden: E.J. Brill.

Ballantyne, W.J. (1961) A biologically-defined exposure scale for the comparative description of rocky shores. *Field Studies* 1(3): 1–19.

Bamber, R.N. (2010) *Sea-Spiders (Pycnogonida) of the North-east Atlantic*, 2nd edn. Synopses of the British Fauna (New Series) 5. Shrewsbury: Field Studies Council.

Bamber, R.N., King, P.E. and Pugh, P.J.A. (2017) Mites and sea spiders (phylum Arthropoda, subclass Acari, Class Pycnogonida) In Hayward, P.J. and Ryland, J.S. (eds),

Handbook of the Marine Fauna of North-West Europe, 2nd edn. Oxford: Oxford University Press.

Björk, M., Axelsson, L. and Beer, S. (2004) Why is *Ulva intestinalis* the only macroalga inhabiting isolated rockpools along the Swedish Atlantic coast? *Marine Ecology Progress Series* 84: 106–116. https://doi.org/10.3354/meps284109

Blaber, S.J.M. (1970) The occurrence of a penis-like outgrowth behind the right tentacle in spent females of *Nucella lapillus* (L.). *Proceedings of the Malacological Society of London* 39: 231–233. https://doi.org/10.1093/oxfordjournals.mollus.a065097

Boissin, E., Chenuil, A. and Féral, J.-P. (2010) Species of the complex *Amphipholis squamata* (Ophiuroidae). In L.G. Harris, S.S. Böttger, C.W. Walker and M.P. Lesser (eds.) *Echinoderms: Proceedings of the 12th International Echinoderm Conference, Durham, New Hampshire, USA, 7–11 August 2006*. Boca Raton, Florida: CRC Press.

Bowen, S., Goodwin, C., Kipling, D. and Picton, B.E. (2018) *Sea Squirts and Sponges of Britain and Ireland*. Plymouth: Wild Nature Press.

Brawley, S.H. (1992) Mesoherbivores. In D.M. John, S.J. Hawkins and J.H. Price (eds), *Plant–Animal Interactions in the Marine Benthos*. Systematics Association Special Volume 46. Oxford: Clarendon Press.

Brodie, J., Walker, R.H., Williamson, C. and Irvine, L. (2013) Epitypification and redescription of *Corallina officinalis* L., the type of the genus, and *C. elongata* Ellis *et* Solander (Corallinales, Rhodophyta). *Cryptogamie, Algologie* 34 (1): 49–56. https://doi.org/10.7872/crya.v34.iss1.2013.49

Bunker, F.StP.D., Brodie, J.A., Maggs, C.A. and Bunker, A.R. (2012) *Seaweeds of Britain and Ireland*. Plymouth: Wild Nature Press.

Bussell, J.A., Lucas, I.A.N. and Seed, R. (2007) Patterns in the invertebrate assemblage associated with *Corallina officinalis* in tide pools. *Journal of the Marine Biological Association of the UK* 87: 383–388. https://doi.org/10.1017/S0025315407055385

Chambers, S.J. and Muir, R.I. (1997) *Polychaetes: British Chrysopetaloidea, Pisionoidea and Aphroditoidea*. Synopses of the British Fauna (New Series) 54. Shrewsbury: Field Studies Council.

Cornelius, P.F.S. (1995) *North-West European Thecate Hydroids and their Medusae*. Parts 1 and 2. Synopses of the British Fauna (New Series) 50. Shrewsbury: Field Studies Council.

Crothers, J. (2012) *Snails on Rocky Sea Shores*. Naturalists' Handbooks 30. Exeter: Pelagic Publishing.

Crump, R.G. and Emson, R.H. (1983) The natural history, life history and ecology of the British species of *Asterina*. *Field Studies* 5: 867–882.

Daniel, M.J. and Boyden, C.R. (1975) Diurnal variations in physico-chemical conditions within intertidal rockpools. *Field Studies* 4: 161–178.

Darrock, D.J. (2011) Phylogeography of two Lusitanian sea stars. Unpublished PhD thesis, Cardiff University.

Davenport, J., Barnett, P.R.O. and McAllen, R.J. (1997) Environmental tolerances of three species of the harpacticoid copepod genus *Tigriopus*. *Journal of the Marine Biological Journal of the UK* 77: 3–16. https://doi.org/10.1017/S0025315400033749

Davenport, J., Healey, A., Casey, N. and Heffron, J.J.A. (2004) Diet-dependent UVAR and UVBR resistance in the high shore harpacticoid copepod *Tigriopus brevicornis*. *Marine Ecology Progress Series* 276: 299–303. https://doi.org/10.3354/meps276299

Emson, R.H. and Crump, R.G. (1984) Comparative studies on the ecology of *Asterina gibbosa* and *A. phylactica* at Lough Ine. *Journal of the Marine Biological Association of the UK* 64: 35–53. https://doi.org/10.1017/S0025315400059622

Emson, R.H. and Whitfield, P.J. (1989) Aspects of the life history of a tide pool population of *Amphipholis squamata* (Ophiuroidea) from South Devon. *Journal of the Marine Biological Association of the UK* 69: 27–41. https://doi.org/10.1017/S0025315400049080

Engel, C.R. and Destombe, C. (2002) Reproductive ecology of an intertidal red seaweed, *Gracilaria gracilis*: influence of high and low tides on fertilization success. *Journal of the Marine Biological Association of the UK* 82: 189–192. https://doi.org/10.1017/S0025315402005349

Fish, J.D. and Fish, S. (2011) *A Student's Guide to the Seashore*. Cambridge: Cambridge University Press.

Fretter, V. and Graham, A. (1962) *British Prosobranch Molluscs*. London: The Ray Society.

Fretter, V. and Manly, R. (1977) Algal associations of *Tricolia pullus* (Linn.), *Lacuna vincta* (Montagu) and *Cerithiopsis tubercularis* (Montagu) with special reference to the settlement of their larvae. *Journal of the Marine Biological Association of the UK* 57: 999–1017. https://doi.org/10.1017/S0025315400026084

George, J.D. and Hartmann-Schröder, G. (1985) *Polychaetes: British Amphinomida, Spintherida and Eunicida*. Synopses of the British Fauna (New Series) 32. Leiden: E.J. Brill.

Gibson, R. and Knight-Jones, E.W. (2017) Flatworms and ribbon worms (Phyla Zenacoelomorpha, Platyhelminthes and Nemertea) In P.J. Hayward and J.S. Ryland (eds), *Handbook of the Marine Fauna of North-West Europe*, 2nd edn. Oxford: Oxford University Press.

Goodwin, C., Picton, B.E., Morrow, C.C. and Dyrynda, P.E.J. (2017) Sponges (phylum Porifera) In P.J. Hayward and J.S. Ryland (eds), *Handbook of the Marine Fauna of North-West Europe*, 2nd edn. Oxford: Oxford University Press.

Green, J. and Macquitty, M. (1987) *Halacarid Mites*. Synopses of the British Fauna (New Series) 36. Leiden: E.J. Brill/W. Backhuys.

Hayward, P.J. (1985) *Ctenostome Bryozoans*. Synopses of the British Fauna (New Series) 33. Leiden: E.J. Brill/W. Backhuys.

Hayward, P.J. (1988) *Animals on Seaweed*. Naturalists' Handbooks 9. Richmond, UK: Richmond Publishing Co. Ltd.

Hayward, P.J. (2004). *Seashore*. New Naturalist 94. London: HarperCollins.

Hayward, P.J. and Ryland, J.S. (1985) *Cyclostome Bryozoans*. Synopses of the British Fauna (New Series) 34. Leiden: E.J. Brill/W. Backhuys.

Hayward, P.J. and Ryland, J.S. (1998) *Cheilostomatous Bryozoa. Part 1. Aeteoidea – Cribrilinoidea*, 2nd edn. Synopses of the British Fauna (New Series) 10. Shrewsbury: Field Studies Council.

Hayward, P.J. and Ryland, J.S. (1999) *Cheilostomatous Bryozoa. Part 2. Hippothooidea – Celleporoidea*, 2nd edn. Synopses of the British Fauna (New Series) 14. Shrewsbury: Field Studies Council.

Hayward, P.J. and Ryland, J.S. (2017) (eds), *Handbook of the Marine Fauna of North-West Europe*, 2nd edn. Oxford: Oxford University Press. https://doi.org/10.1093/acprof:oso/9780199549443.001.0001

Hind, K.K. and Saunders, G.W. (2013) A molecular phylogenetic study of the tribe Corallineae (Corallinales, Rhodophyta) with an assessment of genus-level taxonomic features and description of novel genera. *Journal of Phycology* 49: 103–114. https://doi.org/10.1111/jpy.12019

Hind, K.R., Gabrielson, P.W., Lindstrom, S.C. and Martone, P.T. (2014) Misleading morphologies and the importance of sequencing type specimens for resolving coralline taxonomy (Corallinales, Rhodophyta): *Pachyarthron cretaceum* is *Corallina officinalis*. *Journal of Phycology* 50: 760–764. https://doi.org/10.1111/jpy.12205

Hull, S.L. (1997) Seasonal changes in diversity and abundance of ostracods on four species of intertidal algae with differing structural complexity. *Marine Ecology Progress Series* 161: 71–62. https://doi.org/10.3354/meps161071

Hull, S.L. (1999a) Intertidal ostracod (Crustacea: Ostracoda) abundance and assemblage structure within and between four shores in north-east England. *Journal of the Marine Biological Association of the UK* 79: 1045–1052. https://doi.org/10.1017/S0025315499001289

Hull, S.L. (1999b) Comparison of tidepool phytal ostracod abundance and assemblage structure on three spatial scales. *Marine Ecology Progress Series* 182: 201–208. https://doi.org/10.3354/meps182201

Huys, R., Gee, J.M., Moore, C.G. and Hamond, R. (1996) *Marine and Brackish Water Harpacticoid Copepods*, 2nd edn. Synopses of the British Fauna (New Series) 51. Shrewsbury: Field Studies Council.

Ingle, R.W. (1996) *Shallow Water Crabs*, 2nd edn. Synopses of the British Fauna (New Series) 25. Shrewsbury: Field Studies Council.

Ingle, R.W. and Christiansen, M.E. (2004) *Lobsters, Mud Shrimps and Anomuran Crabs*. Synopses of the British Fauna (New Series) 55. Shrewsbury: Field Studies Council.

Jones, A.M. and Baxter, J.M. (1987) *Molluscs: Caudofoveata, Solenogastres, Polyplacophora and Scaphopoda*. Synopses of the British Fauna (New Series) 37. Leiden: E.J. Brill/W. Backhuys.

Kaiser, M.J, Attrill, M., Jennings, S., Thomas, D.N., Barnes, D.A., Brierley, A.S., Polunin, N.V.C., Raffaelli, D.G. and Williams, P. (2020) *Marine Ecology: Processes, Systems, and Impacts*, 3rd edn. Oxford: Oxford University Press.

Knight-Jones, P. and Knight-Jones, E.W. (1977) Taxonomy and ecology of British Spirorbidae (Polychaeta). *Journal of the Marine Biological Association of the UK* 57: 453–500. https://doi.org/10.1017/S002531540002186X

Knight-Jones, P., Knight-Jones, E.W., Mortimer-Jones, K., Nelson-Smith, A., Schmelz, R.M. and Timm, T. (2017) Annelids (phylum Annelida). In P.J. Hayward and J.S. Ryland (eds), *Handbook of the Marine Fauna of North-West Europe*, 2nd edn. Oxford: Oxford University Press.

Latham, H. (2008) Temperature stress-induced bleaching of the coralline alga *Corallina officinalis*: a role for the enzyme bromoperoxidase. *Bioscience Horizons* 1: 104–113. https://doi.org/10.1093/biohorizons/hzn016

Le Gac, M., Féral, J.-P., Poulin, E., Veyret, M. and Chenuil, A. (2004) Identification of allopatric clades in the cosmopolitan ophiuroid species complex *Amphipholis squamata* (Echinodermata). The end of a paradox? *Marine Ecology Progress Series* 278: 171–178. https://doi.org/10.3354/meps278171

Lincoln, R.J. (1979) *British Marine Amphipoda: Gammaridea*. London: British Museum (Natural History).

Little, C. and Kitching, J.A. (1996). *The Biology of Rocky Shores*. Oxford: Oxford University Press.

Little, C., Williams, G.A. and Trowbridge, C.D. (2009) *The Biology of Rocky Shores*, 2nd edn. Oxford: Oxford University Press.

Littler, M.M. and Kauker, B.J. (1984) Heterotrichy and survival strategies in the red alga *Corallina officinalis* L. *Botanica Marina* 27: 37–44. https://doi.org/10.1515/botm.1984.27.1.37

Lobban, C.S. and Harrison, P.J. (1997) *Seaweed Ecology and Physiology*. Cambridge: Cambridge University Press.

Martins, G.M., Hawkins, S.J., Thompson, R.C. and Jenkins, S.R. (2007) Community structure and functioning in intertidal rock pools: effects of pool size and shore height at different successional stages. *Marine Ecology Progress Series* 329: 43–55. https://doi.org/10.3354/meps329043

McAllen, R. (1999) *Enteromorpha intestinalis* – a refuge for the high-shore rockpool copepod *Tigriopus brevicornis*. *Journal of the Marine Biological Association of the UK* 79: 1125–1126. https://doi.org/10.1017/S0025315499001393

McAllen, R. (2001) Hanging on in there – position maintenance by the high-shore rockpool harpacticoid copepod *Tigriopus brevicornis*. *Journal of Natural History* 35: 1821–1829. https://doi.org/10.1080/00222930110098120

McAllen, R.J., Taylor, A.C. and Davenport, J. (1998) Osmotic and body density response in the harpacticoid copepod *Tigriopus brevicornis* in supralittoral rock pools. *Journal of the Marine Biological Journal of the UK* 78: 1143–1153. https://doi.org/10.1017/S0025315400044386

Naylor, E. and Brandt, A. (2015) *Intertidal Marine Isopods*, 2nd edn. Synopses of the British Fauna (New Series) 3. Shrewsbury: Field Studies Council.

Ó Corcora, T., Davenport, J. and Jansen, M.A.K. (2016) The copepod *Tigriopus brevicornis* (O.F. Müller, 1776) gains UV protection by feeding on UV-acclimated algae. *Journal of Crustacean Biology* 36: 658–660. https://doi.org/10.1163/1937240X-00002462

Oliver, P.G., Holmes, A.M., Killeen, I.J. and Turner, J.A. (2016). *Marine Bivalve Shells of the British Isles*. Available from: http://naturalhistory.museumwales.ac.uk/britishbivalves

Picton, B.E. and Morrow, C.C. (1994) *A Field Guide to the Nudibranchs of the British Isles*. London: Immel Publishing Ltd.

Picton, B.E. and Morrow, C.C. (2016) *Encyclopaedia of Marine Life of Britain and Ireland*. Available at: http://www.habitas.org.uk/marinelife/

Platt, H.M. and Warwick, R.M. (1983) *Free-living Marine Nematodes*, I: *British Enoplids*. Synopses of the British Fauna (New Series) 28. Cambridge: Cambridge University Press.

Platt, H.M. and Warwick, R.M. (1988) *Free-living Marine Nematodes*, II: *British Chromadorids*. Synopses of the British Fauna (New Series) 38. London and Leiden: E.J. Brill/W. Backhuys.

Pleijel, F. and Dales, R.P. (1991) *Polychaetes: British Phyllodocoideans, Typhloscolecoideans and Tomopteroideans*. Synopses of the British Fauna (New Series) 45. Oegstgeest: UBS/W. Backhuys.

Porter, J. (2012) *Seasearch Guide to Bryozoans and Hydroids of Britain and Ireland*. Ross-on-Wye: Marine Conservation Society.

San Martin, G. and Worsfold, T.M. (2015) Guide and keys for the identification of Syllidae (Annelida, Phyllodocida) from the British Isles (reported and expected species). *Zootaxa* 488: 1–29. https://doi.org/10.3897/zookeys.488.9061

Schuchert, P. (2012) *North-West European Athecate Hydroids and their Medusae*. Synopses of the British Fauna (New Series) 59. Telford: Field Studies Council.

Smaldon, G. (1979) *British Coastal Shrimps and Prawns*. Synopses of the British Fauna (New Series) 15. London: Academic Press.

Smaldon, G., Holthuis, L.B. and Fransen, C.H.J.M. (1993) *Coastal Shrimps and Prawns*, 2nd edn. Synopses of the British Fauna (New Series) 15. Shrewsbury: Field Studies Council.

Southgate, T. (1982) The biology of *Barleeia unifasciata* (Gastropoda: Prosobranchia) in red algal turfs in S.W. Ireland. *Journal of the Marine Biological Association of the UK* 62: 461–468. https://doi.org/10.1017/S0025315400057398

Southward, A.J. and Southward, E.C. (1977) Distribution and ecology of the hermit crab *Clibanarius erythropus* in the western channel. *Journal of the Marine Biological Association of the UK* 57: 441–452. https://doi.org/10.1017/S0025315400021858

Southward, A.J. and Southward, E.C. (1988) Disappearance of the warm-water hermit crab *Clibanarius erythropus* from south-west Britain. *Journal of the Marine Biological Association of the UK* 68: 409–412. https://doi.org/10.1017/S0025315400043307

Southward, E.C. and Campbell, A.C. (2006) *Echinoderms*. Synopses of the British Fauna (New Series) 56. Shrewsbury: Field Studies Council.

Tebble, N. (1976) *British Bivalve Seashells: a Handbook for Identification*, 2nd edn. Edinburgh: HMSO.

Thompson, T.E. (1976) *Biology of Opisthobranch Molluscs*, I. London: The Ray Society.

Thompson, T.E. (1988) *Molluscs: Benthic Opisthobranchs*. Synopses of the British Fauna (New Series) 8. London and Leiden: E.J. Brill/W. Backhuys.

Thompson, T.E. and Brown, G.H. (1984) *Biology of Opisthobranch Molluscs*, II. London, The Ray Society.

Walker, R.H., Brodie, J., Russell, S., Irvine, L.M. and Orfanidis, S. (2009) Biodiversity of coralline algae in the northeastern Atlantic including *Corallina caespitosa* sp. nov. (Corallinoideae, Rhodophyta). *Journal of Phycology* 49: 287–297. https://doi.org/10.1111/j.1529-8817.2008.00637.x

Warwick, R.M., Platt, H.M. and Somerfield, P.J. (1998) *Free-living Marine Nematodes*, III: *Monhysterids*. Synopses of the British Fauna (New Series) 53. Shrewsbury: Field Studies Council.

Westheide, W. (2008) *Polychaetes: Interstitial Families*, 2nd edn. Synopses of the British Fauna (New Series) 44. Shrewsbury: Field Studies Council.

White, G.E., Hose, G.C. and Brown, C. (2015). Influence of rock-pool characteristics on the distribution and abundance of inter-tidal fishes. *Marine Ecology* 36 (4): 1332–1344. https://doi.org/10.1111/maec.12232

Wigham, G.D. (1975) The biology and ecology of *Rissoa parva* (da Costa). (Gastropoda: Prosobranchia). *Journal of the Marine Biological Association of the UK* 55: 45–67. https://doi.org/10.1017/S0025315400015745

Wigham, G.D. (2022) *Marine Gastropods 4, Heterobranvhia 1*. Synopses of the British Fauna (New Series) 63. Telford: Field Studies Council.

Wigham, G.D. and Graham, A. (2017a) *Marine Gastropods 1: Patellogastropoda and Vetigastropoda*. Synopses of the British Fauna (New Series) 60. Telford: Field Studies Council.

Wigham, G.D. and Graham, A. (2017b) *Marine Gastropods 2: Littorinimorpha and Other, Unassigned, Caenogastropoda*. Synopses of the British Fauna (New Series) 61. Telford: Field Studies Council.

Wigham, G.D. and Graham, A. (2018) *Marine Gastropods 3: Neogastropoda*. Synopses of the British Fauna (New Series) 62. Telford: Field Studies Council.

Williamson, C.J., Brodie, J., Goss, B., Yallop, M., Lee, S. and Perkins, R. (2014) *Corallina* and *Ellisolandia* (Corallinales, Rhodophyta) photophysiology over daylight tidal emersion: interactions with irradiance, temperature and carbonate chemistry. *Marine Biology* 161: 2951–2068. https://doi.org/10.1007/s00227-014-2485-8

Wood, C. (2013) *Sea Anemones and Corals of Britain and Ireland*, 2nd edn. Plymouth: Wild Nature Press.

Index

References to figures and photographs appear in *italic* type; those in **bold** type refer to tables.

Acanthochitona crinita 108, *108*
Acanthochitona fascicularis 108
Acanthodoris pilosa 127, *127*
Acari (mites) 55–6
Achelia echinata 68, *68*
acidity (pH) 10, 13–14, 22–3, *22*, 154
Acidostoma sarsi 77, *77*
acontia 132, 133, 134
acrorhagi 129, 130, 132
Actinia equina (Beadlet Anemone) *17*, 128, 129, *130*, *140*
Actiniaria (sea anemones) 3, 35, 58, 128–34, *128*
Actinothoe sphyrodeta 135, *135*
Aeolidia papillosa 45
Aetea anguina 64, *64*
Aetea sica 64, *64*
Aiptasia couchii 134, *134*
Alcyonidium gelatinosum 62, *62*
Alcyonidium hirsutum 62, *62*, 68
Alcyonium digitatum 45
algal holdfasts 19, 56, 75, 112–13, 114
algal ostracod communities 52–3
Alvania punctura 104, *104*
Amathia citrina 66, *66*
Amathia gracilis 66, *66*
Amathia imbricata 66, *66*
Amathia lendigera 65, *65*
Amathia pustulosa 66
Ammonicera rota 44, 45, *45*, 103, *103*
Ammothella longioculata 68, *68*
Ammothella longipes 68, *68*
Amphilochus manudens 78, *78*
Amphipholis squamata 30, 47, 48–9, 138, *138*, 158, *159*
amphipods 57, *57*, 75–92
  detritivores/micrograzers 19
  morphology *75*
Ampithoe gammaroides 86, *86*
Ampithoe rubricata 86, *86*
Anemonia viridis (Snakelocks Anemone) 35, *37*, 129, *129*
anomuran porcelain crabs 93
Anoplodactylus petiolatus 69, *69*
Anthopleura ballii 132, *132*
Anthopleura thallia 132
Anthura gracilis 71, *71*
Anurida maritima 56, *56*
Apherusa bispinosa 81, *81*
Apherusa jurinei 81–2, *82*
Apletodon dentatus (Small-headed Clingfish) 145, *145*
Aplysia punctata 122, *122*
Aplysiida 121
Apocorophium acutum 84, *84*
Apohyale prevostii 90, *90*
Arenicola marina (lugworm) 2
Artemia (brine shrimp) 38, *39*
ascidians (sea squirts) 59, 79, 81
Ascophyllum nodosum (Egg Wrack) 15, 27, 107
Asterias rubens 136, 139
Asterina gibbosa (Common Cushion Star) 47–8, 139, *139*
Asterina phylactica (cushion star) 36–7, 47–8, 139, *139*, 158
Asteroidea (starfish) 58, 136, 138–9
Astropecten irregularis 139
Athanas nitescens (Hooded Shrimp) 93, 95, *95*
Aulactinia verrucosa (Gem Anemone) 131–2, *131*

Barleeia unifasciata 42–3, *42*, 106–7, *107*, 155, *158*
barnacles *17*, 127, 152, *153*
Beadlet Anemone (*Actinia equina*) *17*, 128, 129, 130, *130*
bicarbonate ions (HCO_3^-) 10, 13, *13*, 21–2
biological exposure scale (Ballantyne) 152
bivalves 29, 58, 110–13
Bivalvia 110
Black-footed Limpet (*Patella depressa*) 101, *101*
Bladder Wrack (*Fucus vesiculosus*) 15, 22–3, *22*, *23*, 107, 115, *152*
bleaching 10, *13*, 27–8
Blue Mussel (*Mytilus edulis*) 29, 111, *111*

Blue-rayed Limpet (*Patella pellucida*) 2–3, 17, 99–100, *100*
Boreochiton rubra *58*, 109, *109*
Boreostoma variabile 50, *51*
'bottle quadrat' 155–6, *155*
Brachyuran crabs 93
Brania species 119
Branchioma bombyx 118, *118*
brine shrimp (*Artemia* species) 38, *39*
bristle worms (Polychaetes) 19, 58, 114–19, 158
brittle stars (Ophiuroidea) 47–9, 58, 136–7
bromine ions 28
bromoperoxidase 28
bryozoans 59, 60
Butterfish (*Pholis gunnellus*) 35, 148, *148*, 154–5

Calycella syringa 61, *61*
Campecopea hirsuta 72, *72*
Cancer pagurus (Edible Crab) 35, 97–8, *97*
Caprella acanthifera *57*, 91, *91*
Caprella fretensis 92, *92*
Caprella linearis 92, *92*
Caprella septentrionalis 92, *92*
Caprellidae ('skeleton shrimps') 57
Carbon dioxide (CO_2) 10, 13, *13*, 21–2
carbonate ions 13
carbonic acid (H_2CO_3) 10
Carcinus maenas (Shore Crab) 14, 35, *98*
Castalia punctata 120, *120*
Catapaguroides timidus 94, *94*
Celleporella hyalina 63, *63*
Cephalaspidea 121
Ceramium species 18, 32, 34, *34*
Ceramium virgatum *19*, 32, *32*, 49
Cereus pedunculatus (Daisy Anemone) 133, *133*
Cerithiopsis tubercularis (Cerithiopsidae) *34*, 44, *44*, 102, *102*
Cheirocratus sundevallii 83, *83*
China Limpet (*Patella ulyssiponensis*) 2, 17, 101, *101*
chitons 58, *58*, 108–9
Chondrus crispus (Carrageen) 17, *17*, 22–3, *22*, *23*, *34*, 49
Chromadorida 58
Ciliata mustela 147, *147*
Cladophora rupestris 20, *34*, 46, 49
Clibanarius erythropus 94
Clupea harengus (Herring) 141
Clytia hemisphaerica 61, *61*
Codium fragile 20
coenocytic species 46
Collembola 56
commensal cnidarian (*Hydractinia echinata*) *93*
Common Blenny (*Lipophrys pholis*) 35, 149, *149*, 154–5
Common Cushion Star (*Asterina gibbosa*) 47–8, 139, *139*
Common Dog Whelk (*Nucella lapillus*) 43, 99, 101, *101*
Common Goby (*Pomatoschistus microps*) 145, *145*
Common Limpet (*Patella vulgata*) 2, 17, 100, *100*
Common Lobster (*Homarus gammarus*) 35
Common Mussel (*Mytilus edulis*) 29, 111, *111*
Common Periwinkle (*Littorina littorea*) 107, *107*
Common Prawn (*Palaemon serratus*) 35, 93, 96, *96*
Conger Eel (*Conger conger*) 141
Connemara Clingfish (*Lepadogaster candolii*) 146, *146*
Copepoda 56
copepods 19, 37, *37*, 38–9, 56
Corallina caespitosa 25, *25*
Corallina elongata. See Ellisolandia elongata
Corallina officinalis (Coral Weed) *6*, 24–7, 47, 49, 114, 115, 138
  bipartite morphology 25–6
  bleaching 27–8, *27*
  bromoperoxidase activity 28
  circumboreal distribution patterns 24
  fringes in deeper pools 18, *18*
  invertebrate fauna 29–30
  patterns of abundance 51–2
Corallina species
  classification of seaweeds *26*
  density 157
  identifying 15
  occurrence of snail species 34, *34*
Corallina turf 19, 24–30, 40, 155
coralline algae (Corallinales) 10, 17, 24, 29
coralline algal turfs 29, 156–7

Corkwing (*Symphodus melops*) 148–9, *149*
Cornish Clingfish (*Lepadogaster purpurea*) 146, *146*
Coryne pusilla *60*
Coryne species 60
Coryphoblennius galerita (Montagu's Blenny) 150
crabs (decapods) 2, 57, 93–8
Crassicorophium bonellii 85, *85*
Crisia denticulata *65*
Crisia species 65
Crisidia cornuta 64, *64*
crustaceans 29
  amphipods 57
  copepods 56
  isopods 57
  meiofauna 49
  ostracods 56
Ctenolabrus rupestris (Goldsinny) 149
cushion star (*Asterina phylactica*) 36–7, 47–8, 139, *139*, 158
Cyclopterus lumpus (Lumpsucker) 141
Cystoseira 25, 116
Cythere lutea 50, *51*

Dahlia Anemone (*Urticina felina*) 35, *36*, 128, 130, *130*
Daisy Anemone (*Cereus pedunculatus*) 133, *133*
Decapoda (decapods) 2, 19, 57, 93–8
Delesseria sanguinea 30
Dendrodoa grossularia 46
Dexamine spinosa 81, *81*
Diadumene lineata 131, *131*
Diaphoreolis viridis 126, *126*
Diarthrodes nobilis *56*
Diplecogaster bimaculata (Two-spotted Clingfish) 146
diurnal fluctuations 10, *11*
Dog Whelk (*Nucella lapillus*) 43, 99, 101, *101*
Dogfish (*Scyliorhinus canicula*) 141
Doridina 45–6
Doris pseudoargus 46, 127, *127*
Doto coronata 124, *124*
Doto fragilis 45, *45*, *124*
Doto lemchei *124*
Dynamene bidentata 72, *72*

Eatonina fulgida (Cingulopsidae) 42, 106, *106*
echinoderms 3, 36–7, 136–40, 158
Echinogammarus marinus 82, *82*
Echinoidea (sea urchins) 58, 136, 139–40
Edible Crab (*Cancer pagurus*) 35, 97–8, *97*
Edible Sea Urchin (*Echinus esculentus*) 136, *136*, 139
Edwardsiella carnea 132, *132*
Egg Wrack (*Ascophyllum nodosum*) 15, 27, 107
Elasmopus rapax 83, *83*
Electra pilosa 17, 63, *63*
Ellisolandia elongata 25, *25*, 28–9
Elysia viridis 46, *46*, 122, *122*
emersion (period of tidal retreat) 2, 6, 8–10, 14
emersion curves 9, *9*
encapsulated eggs 45
encrusting sponge (*Halichondria panicea*) 44, 59–60, *60*
Endeis 69
Endeis charybdaea 69, *69*
Endeis spinosa 69
Enoplida 58
epilithic animals 59, 108, 155
epiphytic species 2
  Flustrellidra hispida *62*
  gastropods 44, 155
  invertebrates 155
  Patella pellucida 2, 17
Erichthonius brasiliensis 87
Ericthonius brasiliensis 87, *87*
Ericthonius difformis 87
Eubranchus pallidus 126, *126*
euryhaline osmoconformers 38
eurytopic organisms 23
Eusyllis species 119, *119*
extreme high water springs (EHWS) 8–9, *8*
extreme low water mark of spring tides (ELWS) *8*, 15

Facelina auriculata 125, *125*
Fifteen-spined Stickleback (*Spinachia spinachia*) 147, *147*
filamentous red seaweeds 18–19, *19*, 30, 41–2, 54
fish 141–50
Five-bearded Rockling (*Ciliata mustela*) 147, *147*
Flabellina lineata *125*

Flabellina species 45, 125
Flustrellidra hispida 62, *62*, 68, 127
Fucus serratus (Serrated Wrack) 6, 15, 60, 62–3, *63*, 66, 107, 115, *153*
Fucus spiralis 15, *152*
Fucus vesiculosus (Bladder Wrack) 15, 22–3, *22*, *23*, 107, 115, *152*

Gaidropsarus mediterraneus (Shore Rockling) 147, *147*
Galathea squamifera (squat lobster) 97, *97*
gametes 21
gametophytes 21
Gammarella fucicola 83, *83*
Gammarellus angulosus 82, *82*
Gammaropsis maculata 90, *90*
Gasterosteus aculeatus (Three-spined Stickleback) 148, *148*
Gastropoda 99, 121
gastropods 41–7
Gastrosaccus sanctus *57*
Gelidium 18, 33, 34
Gelidium pulchellum 33
Gelidium sesquipedale *34*
Gelidium spinosum 33, *33*
Gem Anemone (*Aulactinia verrucosa*) 131–2, *131*
Gitana sarsi 78, *78*
Gnathia maxillaris 70, *70*
Gobius paganellus (Rock Goby) 144, *144*
Gobiusculus flavescens (Two-spot Goby) 144, *144*
Goldsinny (*Ctenolabrus rupestris*) 149
Gonidoris nodosa 46
Gracilaria gracilis 31
Grantia compressa (Purse Sponge) 60, *60*
Greater Pipefish (*Syngnathus acus*) 142, *142*
Green Sea Urchin (*Psammechinus miliaris*) 140, *140*
green seaweeds 15–16, *16*, 20
Grey Topshell (*Steromphala cineraria*) *102*, 103, *103*
Gutweed. *See Ulva intestinalis* (Gutweed)

Hairy Crab (*Pilumnus hirtellus*) 98, *98*
Halacarellus basteri (marine mite) *55*
Halacaridae 56
Halichondria panicea (encrusting sponge) 44, 59–60, *60*
Halidrys siliquosa (Sea Oak) 16–17
Halurus flosculosus 18, 33, 46
Haplopoma impressum 63, *63*
Harmothoe species 114
Harmothoe extenuata 117, *117*
Harmothoe imbricata 117, *117*
Harmothoe impar 116, *116*
harpacticoid copepods 37–8, *37*
Harpacticoida 56
herbivores 14, 19–20, 46
Hermaea bifida 46, 124, *124*
hermaphroditism 45, 48
hermit crab (*Pagurus bernhardus*) 93–4, *93*, 155
Herring (*Clupea harengus*) 141
Heterobranchia (sea slugs) 45–6, 58, 121–7
heterobranchs 46, 121, 156, 157
Heterocythereis albomaculata *52*, 56
high shore pools 10, 21
Hippolyte inermis 96, *96*
Hippolyte varians 97, *97*
Hirschmannia viridis 50, *51*, 52
holdfasts 19
Homarus gammarus (Common Lobster) 35
honeycomb worms (*Sabellaria*) 151
Hooded Shrimp (*Athanas nitescens*) 93, 95, *95*
Horse Mussel (*Modiolus modiolus*) 110, 111, *111*
humpback prawns (*Hippolyte*) 93
Hyale nilssoni *76*, 91, *91*
hydrogen (H^+) ions 10, 13, *13*
hydrogen peroxide (H_2O_2) 28
hydroids 59–64, *60*
hydroxyl ions (OH^-) 21
Hymeniacidon perlevis 44
hypobromous acid (HBrO) 28

Idotea 70
Idotea baltica 72, *72*
Idotea chelipes *57*, 73, *73*
Idotea emarginata 72, *72*
Idotea granulosa 73, *73*
Idotea neglecta 73, *73*
Idotea pelagica 73, *73*
Iphimedia minuta 79, *79*
Iphimedia obesa 79, *79*
irradiance (sunlight) 10, 28–9

Irus irus 113, *113*
Ischyrocerus anguipes 88, *88*
isopods 19, 57, *57*, 70–4, 158

Jaera albifrons *74*
Jania rubens 25
Janira maculosa 74, *74*
Janua pagenstecheri 116, *116*
Jassa falcata 88, *88*

Kefersteinia cirrata 120, *120*
Kellia suborbicularis 112, *112*
kelp holdfasts 64–5, 72, 74, 77–90, 111, 114, 117–19

Lacuna pallidula 105, *105*
Lacuna parva 106, *106*
Lacuna vincta 33–4, *34*, 106, *106*
Laminaria species 63
Laminaria digitata *6*, 16, 100, *100*
Laminaria hyperborea 16
Lasaea adansoni 112, *112*
Lembos websteri 88, *88*
Leopard-spotted Goby (*Thorogobius ephippiatus*) 144–5
Lepadogaster candolii (Connemara Clingfish) 146, *146*
Lepadogaster purpurea (Cornish Clingfish) 146, *146*
Lepidochitona cinerea 109, *109*
Lepidonotus squamatus 116, *116*
Leptochiton asellus 109, *109*
Leptochiton cancellatus 109, *109*
Leucothoe incisa 79
Leucothoe spinicarpa 79, *79*
Lichina pygmaea 72, 112
Liljeborgia pallida 57, 80, *80*
Limapontia capitata 46, 123, *123*
Limapontia senestra 46, *46*, 123, *123*
limpets (*Patella* species) 2–3, 21, 99, *153*, 155
Lineus ruber *58*
Liparis montagui (Montagu's Sea Snail) 143, *143*
Lipophrys pholis (Common Blenny/ Shanny) 35, 149, *149*, 154–5
lithophilic organisms 25
'lithothamnia' 17
Lithothamnion 17, **26**
Littorina compressa 107
Littorina fabalis 107, *107*
Littorina littorea (Common Periwinkle) 107, *107*
Littorina obtusata 107, *107*
Littorina saxatilis (Rough Periwinkle) 43, 107, *107*
Littorina species 21, 93, 99, 107, 155
Loligo vulgaris (pelagic squid) 3
Lomentaria articulata 18, *19*, 31–2, *31*, 33, *34*, 42
low-shore pools 17
lugworm (*Arenicola marina*) 2
Lumpsucker (*Cyclopterus lumpus*) 141

magnesium chloride 156
major invertebrate animal groups, key to 55–8
marine mite *Halacarellus basteri* *55*
Marshallora adversa 44, 102, *102*
Mastocarpus stellatus 15, *16*, 32, 34, *34*
mean high water mark of neap tides (MHWN) 8–9, *8*, *9*
mean low water mark of spring tides (MLWS) 8–9, *8*, *9*
mean tidal level (MTL) 8, 9, 15
Megalomma vesiculosum 118, *118*
meiofauna 29, 49
meiofaunal ostracods 49–53
Membranipora membranacea 63, *63*
menthol crystals 156
mesoherbivores (mesograzers) 14, 19–20, 46
Metridium dianthus (Plumose Anemone) 131, *131*
Microdeutopus gryllotalpa 89, *89*
Microdeutopus versiculatus 89, *89*
'microgastropods' 45
Microjassa cumbrensis 87, *87*
Microprotopus maculatus 89, *89*
mites (Acari) 55–6
Modiolus barbatus 111, *111*
Modiolus modiolus (Horse Mussel) 110, 111, *111*
modular animals 55, 59
Monhysterida 58
Monocorophium sextonae 85, *85*
Montagu's Blenny (*Coryphoblennius galerita*) 150
Montagu's Sea Snail (*Liparis montagui*) 143, *143*
'mosaic' communities *153*
MTL. *See* mean tidal level
Munna kroyeri *73*

Musculus costulatus 112, *112*
Musculus discors 112, *112*
Musculus marmoratus 111, *111*
mussels *17*, 110, 111, 152
Mysidae 57
mysids (opossum shrimps) 57, *57*
Mytilus edulis (Blue Mussel) 29, 111, *111*

narcotisation 156
'neap tides' *7*, *8*
Necora puber (Velvet Swimming Crab) 35, 98, *98*
Nematoda 58
nematodes (round worms) 58, *58*
Nemertea 58
nemerteans (ribbon worms) 58
Neoamphitrite 114
Neoamphitrite edwardsii 117
Neodexiospira pseudocorrugata *115*, 116
Nereis pelagica 118, *118*
Nerophis lumbriciformis (Worm Pipefish) 35, 143, *143*
Nerophis ophidion (Straight-nosed Pipefish) 143, *143*
Nicolea species 114
Nicolea venustula 117, *117*
Nicolea zostericola 117
Nototropis swammerdami 80, *80*
Nucella lapillus (Common Dog Whelk) 43, 99, 101, *101*
Nudibranchia (nudibranchs) 45–6, 121
Nymphon brevirostre 67–8, *67*, *68*
Nymphon gracile 67, *67*

Odontosyllis species 119, *119*
Omalogyra atomus 44, 45, 103, *103*
Omalogyridae 44–5
Onchidella celtica 123, *123*
Onchidoris bilamellata 127, *127*
Onchidoris muricata 46, 127, *127*
Onoba semicostata 29, 105, *105*
Ophiocomina nigra 137, *137*
Ophiopholis aculeata 137, *137*
Ophiothrix fragilis 138, *138*
Ophiuroidea (brittle stars) 47–9, 58, 136–7
opisthobranchs ('sea slugs') 121
opossum shrimps (mysids) 57, *57*
Orchomene humilis 77, *77*
Orthopyxis integra 61, *61*
Osmundea pinnatifida 26, 34, *34*, 42
ostracods 49–53, 56
densities 49–51, *50*, *51*
species 51–2, *52*

paddleworms (Phyllodocidae) 114, 120, *120*
Pagurus bernhardus (hermit crab) 93–4, *93*, 155
Painted Goby (*Pomatoschistus pictus*) 145, *145*
Palaemon elegans 35, 93, 95, *95*
Palaemon longirostris 96, *96*
Palaemon serratus (Common Prawn) 35, 93, 96, *96*
Palaemonetes varians 95, *95*
Parablennius gattorugine (Tompot Blenny) 150
Paracentrotus lividus 140, *140*
Parajassa pelagica 87, *87*
Patella depressa (Black-footed Limpet) 101, *101*
Patella species (limpets) 2–3, 21, 99, *153*, 155
Patella ulyssiponensis (China Limpet) 2, 17, 101, *101*
Patella vulgata (Common Limpet) 2, 17, 100, *100*
pelagic squid (*Loligo vulgaris*) 3
Perinereis cultrifera 119, *119*
pH meters 154
pH values. *See* acidity (pH)
Pheasant Topshell *Tricolia pullus* 34, *34*, 44, *44*, 103, *103*
Pholis gunnellus (Butterfish) 35, 148, *148*, 154–5
Phorcus lineatus (Toothed Topshell) 99, 103, *103*
photo-acclimation 10
photo-inhibition 10, 20
photosynthesis 10, 13–14, 22, 28–9
Phoxichilidium femoratum 69
Phtisica marina 91, *91*
Phyllodocidae (paddleworms) 114, 120, *120*
Pilumnus hirtellus (Hairy Crab) 98, *98*
pipefish (Syngnathidae) 35, 142–3
Pisidia longicornis 97, *97*
Placida dendritica 46, *46*, 124, *124*
Platynereis dumerilii 118, *118*
Pleurobranchida 121
Plumose Anemone (*Metridium dianthus*) 131, *131*

Plumularia setacea 61, *61*
Podocerus variegatus 84, *84*
Polycera quadrilineata 126, *126*
Polychaetes (bristle worms) 19, 58, 114–19, 158
Pomatoschistus microps (Common Goby) 145, *145*
Pomatoschistus pictus 145, *145*
Porcelain Crab (*Pisidia longicornis*) 97, *97*
Porcellana platycheles 97
prawns 35, 57, 94, *94*
predicted tidal curves *8*
preserving soft-bodied specimens 156
propylene phenoxetol 156
Psammechinus miliaris (Green Sea Urchin) 140, *140*
Pseudoprotella phasma 91, *91*
Purple Topshell (*Steromphala umbilicalis*) 102, *102*
Purse Sponge (*Grantia compressa*) 60, *60*
Pycnogonida (sea spiders) 55, 67–9, *67*

quadrats 155–6, *155*

ragworms (*Platynereis*, *Nereis*) 114
red algae/seaweeds 18, 30–1, *153*
red algal fringe 30–4, 114
red algal turfs 48, 56, 67, 156–7
Rhodymenia pseudopalmata 30
ribbon worms (nemerteans) 58
Rissoa guerinii 105, *105*
Rissoa lilacina 104, *104*
Rissoa membranacea 105, *105*
Rissoa parva 29, 41, *41*, 42, 105, *105*, 155
Rissoella (Rissoellidae) 44
Rissoella diaphana 44–5, 106, *106*
Rissoella opalina 106, *106*
rissoids (Rissoidae) 41
Rock Goby (*Gobius paganellus*) 144, *144*
rock limpets 2–3
rock-pool animals
  density and diversity 35
  within the fringe 40–1
rock-pool ecology 14
rock-pool habitats 3, 151–9
rock pools
  defining 6
  studying and measuring 154

Rough Periwinkle (*Littorina saxatilis*) 43, 107, *107*
round worms (nematodes) 58, *58*
Runcina coronata 46–7, 122, *122*
Runcinida 121

Sabellaria (honeycomb worms) 151
Sacoglossa (sea slugs) 46, 121
Sagartia elegans 35, *36*, 128, 133–4, *133*, 135
Sagartia ornata 134
Sagartia troglodytes 134
salinity 8–10, *12*, 154
Salvatoria species 119
Sarsia eximia 60, *60*
scale worms 114
Scruparia ambigua *64*
Scruparia species 64
Scyliorhinus canicula (Dogfish) 141
sea anemones (Actiniaria) 3, 35, 58, 128–34, *128*
Sea Lettuce (*Ulva lactuca*) *6*, 16, *16*, 20–1
Sea Oak (*Halidrys siliquosa*) 16–17
sea slugs (Heterobranchia) 45–6, 58, 121–7
sea spiders (Pycnogonida) 55, 67–9, *67*
sea squirts (ascidians) 59, *79*, 81
sea urchins (Echinoidea) 58, 136, 139–40
seasonal variations 10, 35, 154
seaweed zones 15
seaweeds 15–34
  changing chemical environment 14
  functional-form classification 26, **26**
  photosynthesis 21–2
  upper tidal limit 15
Serrated Wrack (*Fucus serratus*) 6, 15, 60, 62–3, *63*, 66, 107, 115, *153*
sessile animals 2, 55, 59–66
Shanny (*Lipophrys pholis*) 35, 149, *149*, 154–5
sheet-encrusting corallines 15, 17, *17*, 108
shelled gastropods 99–107
Shore Crab (*Carcinus maenas*) 14, 35, *98*
Shore Rockling (*Gaidropsarus mediterraneus*) 147, *147*
shrimps 57
Simpson's Index 157, **157**

'skeleton shrimps' (Caprellidae) 57
Skenea serpuloides (Skeneidae) 44–5, 104, *104*
Skeneopsis planorbis (Skeneopsidae) 44, 45, 104, *104*
Small-headed Clingfish (*Apletodon dentatus*) 145, *145*
Snakelocks Anemone (*Anemonia viridis*) 35, *37*, 129, *129*
sodium concentrations 38, *39*
specimen storage 156
Sphaerosyllis species 119, *119*
Sphenia binghami 113, *113*
Spinachia spinachia (Fifteen-spined Stickleback) 147, *147*
Spiral Wrack (*Fucus spiralis*) 15, *152*
Spirorbis corallinae (tubeworm) 29, 30, 55, 115, *115*
Spirorbis inornatus 116, *116*
Spirorbis rupestris 115, *115*
Spirorbis spirorbis 115, *115*
sponges 44, 55, 59–60
Spotted Cowrie (*Trivia monacha*) 101, *101*
Sprat (*Sprattus sprattus*) 141
spring tides 7, *8*
squat lobster (*Galathea squamifera*) 97, *97*
standard tidal levels 7, *8*
starfish (Asteroidea) 58, 136, 138–9
Stenothoe marina *78*, *78*
Stenothoe monoculoides *76*, 78, *78*
Steromphala cineraria (Grey Topshell) *102*, 103, *103*
Steromphala umbilicalis (Purple Topshell) 102, *102*
Steromphalus (topshells) 93, 99
Straight-nosed Pipefish (*Nerophis ophidion*) 143, *143*
summer daily temperature ranges 9
Sunamphitoe pelagica 90, *90*
sunlight (irradiance) 10, 28–9
supralittoral rock pools *12*, *13*, 37–9
Sycon ciliatum 60, *60*
Syllidae species 114, 158
Syllis species 119, *119*
Symphodus melops (Corkwing) 148–9, *149*
Syngnathidae (pipefish) 142–3
Syngnathus acus (Greater Pipefish) 142, *142*
Syngnathus rostellatus 142
Syngnathus typhle 142
Synisoma acuminatum 72, *72*

Tectura virginea (White Tortoiseshell Limpet) 100, *100*
Tellina tenuis (thin-shelled bivalve) 2
temperature fluctuations 8, 9–10, 154
Tergipedidae 125
Testudinalia testudinalis (Tortoiseshell Limpet) 100, *100*
Thorogobius ephippiatus (Leopard-spotted Goby) 144–5
Three-spined Stickleback (*Gasterosteus aculeatus*) 148, *148*
tidal cycles 7–8
tidal emersion curves 9, *9*
tidal levels 7, 9–10, 151–2
tidal predictions 9, 152
tidal retreat (emersion) 2, 6, 8–10, 14
Tigriopus brevicornis 23, 36–40, *37*, 56–7
  sodium concentrations 38, *39*
  temperature and salinity 38
  ultraviolet radiation 38–40, *40*
Tompot Blenny (*Parablennius gattorugine*) 150
Tonicella marmorea 109
Toothed Topshell (*Phorcus lineatus*) 99, 103, *103*
topshells (*Steromphalus* species) 93, 99
Tortoiseshell Limpet (*Testudinalia testudinalis*) 100, *100*
Tricolia pullus (Pheasant Topshell) 34, *34*, 44, *44*, 103, *103*
Tritaeta gibbosa *80*, 81
Tritonia hombergii 45
Trivia monacha (Spotted Cowrie) 101, *101*
Tryphosa nana *77*, *77*
tubeworm (*Spirorbis corallinae*) 29, 30, 55, 115, *115*
Turtonia minuta 113, *113*
Two-spot Goby (*Gobiusculus flavescens*) 144, *144*
Two-spotted Clingfish (*Diplecogaster bimaculata*) 146

ultraviolet radiation 38–40
Ulva species 30–1
Ulva intestinalis (Gutweed) 10, *13*, 16, *16*, 20–3, 38
  intertidal distributions 20–1

organic carbon concentrations 23, *23*
photosynthesis 22–3
Ulva lactuca (Sea Lettuce) *6*, 16, *16*, 20–1
upper-shore pools 10, *17*, 28–9, 31, 99, *153*
Urticina felina (Dahlia Anemone) 35, *36*, 128, 130, *130*

veliger larvae 41, 44, 46
Velvet Swimming Crab (*Necora puber*) 35, 98, *98*
Viviparous Blenny (*Zoarces viviparus*) 148, *148*

Walkeria uva 29, 65, *65*
water temperature 8–10, 154
water volume calculations 154
wave-cut platforms 5–6, *5*
wave exposure 152, *153*
White Tortoiseshell Limpet (*Tectura virginea*) 100, *100*
'whitebait' 141
winkles (*Littorina*) 21, 93, 99, 155
Worm Pipefish (*Nerophis lumbriciformis*) 35, 143, *143*

Xestoleberis aurantia 50, *51*

Zoarces viviparus (Viviparous Blenny) 148, *148*
zonation patterns 15